RURAL ENERGY FOR SUSTAINABLE DEVELOPMENT
Technology and Environmental Issues

About the Editors

Pradeep Chaturvedi is a well known expert in the Rural Energy Sector. His expertise lies in project planning, implementation and management with emphasis on the South Asia Region. A number of pioneering projects and studies of national and international significance have been conducted under his guidance. His involvement with the FAO's Regional Wood Energy Development Programme, based at Bangkok, has given him exposure to biomass use for energy in the South Asia Region.

Mr. Chaturvedi is the Vice-Chairman of the UNESCO sponsored World Renewable Energy Network based in UK. He is President of the Bio Energy Society of India, and Secretary General of the Indian Association for the Advancement of Science. He is also a member of the Rural Energy Study Committee of the World Energy Council.

Mr. Chaturvedi is a graduate in Engineering. He is a Fellow of the Institution of Engineers (India) and Fellow of the Institute of Directors. He is also the recipient of FAO's World Food Day Medal, 1991.

Mr. Chaturvedi has edited and authored 24 books related to Sustainable Development, Rural Energy and Energy Conservation, and has contributed papers at a number of national and international conferences.

RURAL ENERGY FOR SUSTAINABLE DEVELOPMENT
Technology and Environmental Issues

PRADEEP CHATURVEDI

Published by

CONCEPT PUBLISHING COMPANY, NEW DELHI-59

for

INDIAN ASSOCIATION FOR THE ADVANCEMENT OF SCIENCE
NEW DELHI

ISBN 81-7022-721-6

First Published 1998

Published and Printed by

Ashok Kumar Mittal
Concept Publishing Company
A/15-16, Commercial Block, Mohan Garden,
NEW DELHI - 110059 (India)

Phones: 5648039, 5649024
Fax: (011) 5648053

Dedicated to

My Friend

Prof. W.S. Hulscher

Chief Technical Adviser
FAO's Regional Wood Energy Development Programme

in recognition for his

deep concern for energy supply to the millions in the rural areas of the Developing Countries; creating Human Capital by training experts from the Developing Countries (at the University of Twente) in planning and implementing energy projects; and leading a Team of Experts at FAO's Regional Wood Energy Development Programme for Asia to encourage member countries to plan for sustainable woodfuel supply as the major source of energy in the rural areas.

ACKNOWLEDGEMENTS

I wish to gratefully acknowledge the help and inputs provided by different experts at different times of writing the book. Availability of authentic information on 7 countries of the South Asia Region could be possible, mainly due to the assistance provided by the National Member Committees of the World Energy Council.

I wish to express my gratitude to Dr. I. Natarajan for providing me inputs about fuelwood consumption pattern.

I am grateful to Prof. W.S. Hulscher, Chief Technical Advisor, FAO's Regional, Wood Energy Development Programme in Asia for various inputs on biomass fuels. My thanks to him also for allowing me to reproduce the Executive Summary of the Regional Study on Wood Energy—Today and Tomorrow in Asia, published by RWEDP as field document No. 50.

I also wish to thank the Indian Association for the Advancement of Science for permitting me to reproduce Dr. Atma Ram Memorial Lecture on "Biomass Energy: A Modern and Environmentally Friendly Fuel for the Future" by Prof. David O. Hall & J.I. House. These two documents will provide additional authentic information for the readers of the book.

Mr. Ashok Mittal, Managing Director, Concept Publishing Co. has been the moving force in bringing out a quality publication at such a short notice, and he deserves my gratitude.

New Delhi
March 17, 1998

PRADEEP CHATURVEDI

CONTENTS

INTRODUCTION

The South Asia Region has 7 countries, viz, India, Pakistan, Nepal, Bangladesh, Bhutan, Sri Lanka and Maldives. These countries have a wide variety of geographic and topographic locations. The region has over 75% of its population living in the rural areas and a substantial number facing acute poverty. Over 1.2 billion people inhabited the South Asia Region in 1994, thereby representing almost 20% of the global population. This will rise to about 1.4 billion in the year 2000, with 80% of them residing in rural areas. The change in rural population in these countries over the last 34 years (1960-94) has not shown any major shift. The rate of urbanisation now taking place indicate that a substantial population will shift from rural areas to major cities. It is expected that major cities with population over 750,000 will be a serious problem as they will be inhabiting significantly high populations with high population densities. This has also resulted in urban slums, created by migratory population. The immensity of the problem can be realised from the fact that underweight children under the age of 5, have been reported to be 50% in South Asia during 1990-96. There are 380 million peoples, who are illiterate, 280 million lack elementary sanitation, and 300 million live in absolute poverty.

At the 16th Congress, at Tokyo in 1995, the World Energy Council had expressed concern for the 2 billion people without access to commercial sources of energy in the developing countries. Of this, almost 50% live in South Asia. The use of primary commercial energy in South Asia increased at an annual rate of 6.8% between 1971 and 1993, but its reach was to a small number of population. Most of the rural masses depend on traditional biomass fuels.

Traditional energy sources have been central to basic nutrition and health needs of the rural poor. The low income households tend to rely on significantly different set of energy carrier than the rich. It is observed, they are still dependent on agro-residues, twigs and branches of the tree which are available free to them. A recent cross country data of 114 countries, presented by UNDP shows linkage between energy consumption and distribution of income. Second, the mix of energy carriers, varies with income and its distribution. Whereas in the rural areas, in India, with larger income transfers due to increasing cereal prices, there is a shift from traditional biomass fuels to kerosene and LPG, other countries are still dependent on traditional fuels.

Fuelwood has been the most sustainable traditional source of energy for the rural areas in South Asia. Fuelwood has been observed to supply around 90% of the domestic energy needs in the rural areas. In addition the rural industries are also dependent on the fuelwood supply only.

FAO's Regional Wood Energy Development Programme in Asia carried out a Regional Study on Wood Energy - Today and Tomorrow has observed that for all countries, the consumption of biomass fuel is increasing, whereas the share of biomass energy in the total energy consumption is declining for most countries and is stable for some.

The fuelwood 'gap theory' often quoted in the past was based on the belief that most, if not all, fuelwood originated from forests. The gap between demand and supply was then used to calculate how long it would take before all the forest wood disappear due to fuelwood use. However, 10-15 years of in-depth studies have shown that non-forest areas supply the considerable amount of fuelwood. In fact, evidence shows that in many countries often over 50% of fuelwood is derived from non-forest areas. Latter include village land, agricultural land, agricultural crop plantation, homesteads and trees along roads.

The fuelwood production has deep linkages with policy issues in forestry, agriculture, energy, rural development and environment. The consumption of fuelwood is not likely to reduce in years to come. Due to its beneficial advantage for the global environment fuelwood is expected to be a preferable energy source. Fuelwood supply also implies labour for growing, harvesting, processing, wholesaling, transporting and retailing the products. Per unit of energy, the labour involved in these fuelwood business is about 20 times larger than for kerosene. Fuelwood business is also the main source of income for about 10% of rural households.

Whereas the growth of fuelwood, in India, presently is on public lands and through government efforts, it is expected that the future growth in fuelwood supply will be on market forces and through commercial channels. It is not to say that the existing channels will be totally stopped. The sustainable aggregated potential supply of fuelwood is expected to outweigh, aggregated consumption. The fuelwood gaps will really be taken care of by the localised fuelwood supply. At present the sustainable potential supply of fuelwood from agricultural lands, more or less meet the consumption. While the overall supply/demand balance of wood and other biomass fuels will not be a major concern, their distribution may become such a constraint.

The future fuelwood supply to the South Asia Region will depend on the steps taken now in the field of R&D for higher production of biomass for food, fuel and fodder, so that the available land mass can meet all the demands in three sectors. The key to success lies in evolving complementary roles and developing technique that can ensure sustainable biomass growth to the year 2010 and beyond. These techniques should enable the entrepreneurs to grow fuelwood on a self-sustaining basis and wherever possible on waste and marginal lands.

The objective of compiling this publication is to bring at one place the synthesis of expert opinions; and let the policymakers and users take appropriate decisions for uninterrupted supplies of biomass for fuel in developing countries.

To provide relevant authentic information of recent study and the expert opinion, an Annexure has been added. It has a Paper on Dr. Atma Ram Memorial on "Biomass Energy: A Modern and Environmentally Friendly Fuel for the Future" by Dr. David O. Hall, delivered on February 17, 1995 under the Indian Association for the Advancement of Science; and the Executive Summary of the "Regional Study on Wood Energy: Today and Tomorrow in Asia" concluded by the FAO's Wood Energy Development Programme in Asia. Field Document No. 50.

I am extremely grateful to IAAS and RWEDP for giving me permission to include these two documents as Annexure and thereby allow wider reach for the views incorporated therein.

I have the privilege of Dedicating this Book to my friend Prof. W.S. Hulscher, the Chief Technical Adviser, RWEDP for his untiring efforts to assist governments in 16 participating countries to plan systematic approach for ensuring fuelwood supplies to meet the fuel needs of the masses in Asia. I wish to express my gratitude to Prof. Hulscher for offering me an opportunity to express my appreciation to him and his team at RWEDP.

I hope the readers will find the Book useful.

New Delhi **PRADEEP CHATURVEDI**
March 17, 1998

1

FUELWOOD FOR RURAL ENERGY IN SOUTH ASIA

Introduction

Rural Energy Supply in South Asia has been mainly from fuelwood, agro residues and tree twigs. The region is worst affected by poverty and has the highest incidence of malnutrition. Being the most vulnerable region, its economy has not grown as fast as is necessary for its economic development. One major hurdle for the same has been the short supply of commercial sources of energy. Most of the rural energy systems are of a small scale and not covered by any single rural energy policy. Rural areas have their activities covered under rural development, agriculture, water supply, forestry, village and cottage industries etc. The individual policies in each country have a large bearing on the evolution of rural energy systems, being mostly dependent on renewable sources of energy and no country barring India having any policy on renewable sources of energy, the supply of rural energy has been mostly governed under policy for forestry and that for agriculture. Another important and significant aspect is the lack of sufficient data on rural energy supply or demand. In some countries such a data is now being developed.

Regional Wood Energy Development Programme in Asia covers 16 major countries in South and South-East Asia. It has come out with a Regional Study on Wood Energy—Today and Tomorrow. It indicates that the developing countries in Asia inhabits three-quarters of world's woodfuel users, but have only one quarter of the forest cover in the South. That sounds dramatic, but it is not. It has been found that about two-thirds of all woodfuels originate from non-forest land. The implications of these important findings are still to be absorbed by many policy makers. Major implications are that woodfuel consumption is not a general or main cause of deforestation, and then woodfuel consumption will remain, whether or not there are forests. The future of Asia's tropical forests and the problems of woodfuel users are not as closely linked as is often assumed.

The study also emphasises that in the 1970s and 1980s it was generally assumed that all woodfuels originate from public forest land. The lead to the 'fuelwood gap theory' from which it was concluded that non-sustainable yield were taken from forests to meet wood energy demands. The theory resulted in the assumption that woodfuel use was a root cause of deforestation. But now ample evidence exists to prove that the theory is false and that, except for localised areas, woodfuel use is not a main cause of deforestation. The indications are that a major cause of deforestation is the ongoing conversion of forest land into other land uses, particularly agriculture. This is generally carried out by planned forest clearing or results from gradual processes of forest encroachment.

The rural energy supply system is essentially a system built on the natural resource base in the immediate vicinity of human settlement. Over the years, in many developing countries, due to expanding population and their resultant increase in their energy needs, this system has drawn upon local resources at a rate much faster than their replenishment rate. These countries are now faced with the environmental stress of varying degrees of severity, and are hard pressed to shift the energy demands of their rural populations away from the self strangulating process of eroding the very resource base that provides them sustenance.

There is an increasing need for energy in activities that enhance economic productivity, such as agricultural mechanisation, irrigation, the use of fertilizers, the expansion of rural industries and other infrastructure development. The social needs are also increasing at a fast pace. Ideally, both sets of needs have to be met in efficient, cost effective and environmentally sustainable manner. This is possible only by achieving a transition from inefficient traditional energy sources and technologies to more efficient modern sources of energy. Developing countries for the South-Asia region are struggling to achieve this objective. While the general need and the thrust of the efforts to promote rural energy development is fairly similar, the progress achieved by each country varies according to its natural resource endowment, extent of policy commitment, institutional effectiveness and capacity to relate the energy needs at the grass-roots level.

Socio-economic Situation in South Asia

The average rural population of all the South Asian countries was 75% of national population in 1994-95. The total population of this region was about 1.2 billion. Of this population, one-half have no access to health services and almost two-thirds went without sanitation, nearly one out of four adults are illiterate and six out of every 100 children suffered from malnutrition.

South Asia covers over one-fifth of the humanity and this population of 1.2 billion is growing rapidly. More people are borne in South Asia every year (27 million) than that the total population of all the Scandinavian countries (Denmark, Finland, Norway and Sweden). As a matter of fact the South Asian population exceeds the total population of 50 of the smaller UN Member countries.

In the global context, the South Asia region has begun to lag behind all other regions of the world, both in its income and in human development levels. South Asia is by now the poorest region in the world—its per capita income of US $ 309 is much below US $ 555 for Sub-Saharan Africa and is only one-third of US $ 970 for all developing countries.

Its share in world population is 22%, and contains nearly 40% of the world's absolute poor.

Adult literacy are of 48% large behind 58% rate achieved in Sub-Saharan Africa and 98% in East Asia (Excluding China). South Asia also experienced an increase in its share of the world's illiterate adult population, from 32% in 1960 to 46% in 1993, demonstrating that adult literacy in this region increased at a much slower pace than in other regions and also that the school education was not on a desired level.

The extent of human deprivation in South Asia can be summed up as follows:

(i) over 500 million people live in absolute poverty
(ii) 260 million people lack access to even rudimentary health facilities.
(iii) 337 million people are without safe drinking water.
(iv) 830 million people have no access to basic sanitation.
(v) 396 million adults are unable to read and write.
(vi) 600 million people have no proper shelter.

The potential for economic growth, employment and income generation in the rural areas is centred around the state of agricultural development, which is the source of primary, food production and the base for allied activities, such as cottage and small scale industry, and off-farm village enterprises. In the face of growing population and declining human-land ratio, the enhancement of agricultural productivity and other income generating activities through the introduction of modern technologies and inputs is, therefore, a major policy objective in all countries. This is accompanied by efforts to improve the quality of life in the rural areas to infrastructure development, such as access roads and communication facilities, and the provision of social amenities in the form of health care, education and water supply, among others.

Role of Energy in Rural Development

The priority is to provide a minimum amount of energy to meet the basic needs of the people, particularly the poor and the economically disadvantaged group, irrespective of their ability to bear the cost of supply and delivery from a longer term view point, it is necessary to accelerate the modernisation progress of the rural energy sector to facilitate and sustain productive economic activities in the rural areas in order to augment the income levels of the people and to strengthen their capacity to self-reliance.

Asia Pacific Development Centre, APDC has observed that the twin focus of rural energy development often tend to overlap and may seem to conflict with one another at times. For example, the welfare objective signified by government subsidies in the price of rural electricity, of commonly used fuels, such as kerosene and diesel, is often unrealised because such lower price energy sources tend to be diverted from the people for whom they are intended and, instead, end up with the more affluent segments of the rural society, who could afford the real cost of energy supply. This leads to the inefficient allocation of fuels and their wasteful use, and affects the financial viability of the energy supply in the district. A reduction or removal of pricing subsidies to overcome this problem is, however, rendered complex because of its potential effects on the poor and low income population. The partial withdrawal of kerosene price subsidy in Sri Lanka and Myanmar, for instance, has forced people in some rural areas to fall back upon fuelwood. Both these countries suffered from deforestation and such a move would have serious environmental implications for them in the long run.

FAO's Regional Wood Energy Development Programme,

RWEDP, has observed that a critical issue in rural energy development is the non-monetized nature of many aspects of the rural economy, in particular the bulk of energy supplies which is in the form of fuel wood and biomass. Limited cash as rural people have, could be applied to a variety of uses, among which energy, (which has traditionally considered a free resource), have a lower priority than others. At the same time, moving beyond basic need satisfaction towards higher level of economic productivity would mean the introduction into rural societies of modern energy sources, all of which carry a cash price. While the ability and willingness of the rural people to make the transition from traditional to modern energy sources is contingent upon their levels of income, their prospects of achieving high income levels, are, in turn constrained by the extent to which such a transition is achieved.

Recent Trends in Rural Energy Development

Rural economies continue to rely heavily on traditional sources of energy based on primitive technology, and the supply of modern energy sources for rural activities remains low. The present pattern of rural energy consumption have several implications, of which APDC has observed the following are significant from the rural development view point:

(a) The direct combustion of fuelwood using inefficient technologies and appliances results in a wastage of wood resources and increases the problem of deforestation and its downstream effect on agriculture.
(b) The use of crop residue and animal waste, as fuel, depletes soil quality and agricultural productity since these resources have alternative applications as organic fertilizers.
(c) The time spent on collecting and using traditional fuels

represents a significant loss of human availability for self development and productive economic activity.

(d) The lack of modern energy limits the opportunities to achieve major structural changes in the rural economies that could lead to enhanced income generating activities and the elimination of poverty.

(e) In the absence of appropriate energy alternative, the rural people are deprived of basic social amenities, such as health services, sanitation, water supply and education.

The above mentioned issues have been commonly acknowledged and their significance has been underlined in recent years due to changes that have taken place in the mainstream rural development. The search for strategies for moving towards better quality fuel is based on the awareness of several realities and convictions. Some of these may be as listed below:

(a) First, during the next decade, and quite possibly beyond, the chances of completely eliminating the dependence of the rural people on traditional energy sources, especially fuelwood, are slim. One can at best offer to reduce this dependence by (i) increasing the supply of conventional commercial energy and/or (ii) introducing decentralized alternative to centralised supplies.

(b) Second, the use of traditional energy sources is not undesirable in itself. In the manner in which these are currently used that needs to be improved. Thus, it is more a matter of technology rather than of resource, as indicated by array of prevailing an emergent advanced technologies that are capable of delivering efficient and economic energy produce from traditional energy sources (biogas, methanol, ethanol and electricity from the gasification of wood or rice husk to name some)

(c) Third, any transition from traditional energy to modern energy should be consistent with the structural changes that are, and would be, taking place in rural economies. One such change that has taken place in all the countries is the gradual expansion of rural industries and enterprises. Another change observed is the increasing energy intensity of agricultural activities due to mechanization and petro-chemical inputs. A more profound change has been the transformation of erstwhile rural areas into semi-urban areas or townships due to a variety of reasons. It is essential to recognize that many rural areas will not always remain rural, and even if they do remain rural, the pattern and volume of energy demand would shift in a manner requiring greater amount of modern energy supply. Future energy systems must, therefore, be capable of not only delivering more convenient energy, but they must also be able to cope up with a substantial scaling up of the demand.

(d) Fourth, in view of the above, efforts to meet the basic energy needs of the people is to be seen as one part of rural energy development, but not all of it. A parallel set of efforts, based on centralized supply system or decentralized system capable of higher output, would be necessary to sustain the economic development of the rural areas. To that extent, a future rural energy development strategy would rely on a combination of centralised and decentralised systems, the share of which would increase steadily in proportion to the declining share of the traditional energy in the supply mix.

In a survey conducted by APDC in 1993 an effort was made to project the overall energy consumption and supply patterns in South Asia. The consumption patterns offered following peculiar characteristics:

(a) Households are the major consumers of energy, their share of the aggregate direct energy consumption averaging about 90%. Most of this is in the form of traditional energy sources for cooking.

(b) Major use of commercial energy will be kerosene and electricity, in the households for lighting which forms 10% of the total consumption.

(c) Agricultural activities consume 5-15% of the aggregate, mainly in the form of commercial energy. Of this, slightly more than half constitute indirect energy inputs by way of chemical inputs and the remainder is for operating mechanical equipment and irrigation pump sets.

(d) Energy consumption of rural industries amounts to about 5% of the aggregate in most countries. Fuelwood and agricultural residue constitute the principal sources of supply for rural industries.

(e) Use of electricity in the rural areas remains low in most countries. The lowest rate of access of 2% is observed in Nepal.

(f) Average share of commercial sources of national energy consumption are 45 and 55% respectively.

(g) Energy supplied by modern decentralized renewable energy systems in all the countries, form less than one percent.

RURAL ENERGY POLICIES AND STRATEGIES

APDC and RWEDP have tried to synthesise the steps taken for rural energy policies.

Evolution of Rural Energy Development Policies

Policy interest in rural energy development can be traced back to the oil price crises of the 1970s when the developing countries of the region were faced with serious oil shortages and, thus, began the drive towards fuel diversification and energy self-reliance. Although, at that time, the consumption of petroleum products in rural areas was a relatively small portion of the overall national consumption, the focus on rural energy issues was derived from the following factors.

1. During the early 1980s, many developing countries embarked on energy master plans or similar exercises that attempted, for the first time, a comprehensive inventory of the national energy situation and an assessment of energy demand-supply patterns. A key outcome of this effort was the discovery of the magnitude of energy demand in the rural areas and the role of traditional energy sources as the principal supply sources catering for rural energy needs. Although the transition from traditional to modern energy sources had not occurred to any significant extent in most countries, the fact that a substantial part of such transition was taking place involved the greater use of oil products was a cause for concern.
2. Due to the overwhelming proportion of the rural population in a number of countries, such as Bangladesh, Myanmar, Nepal and Vietnam, it was realised that energy development in these countries was almost synonymous with rural energy development. Although greater attention was paid to energy development for the modern sectors of the economy in other countries, national concerns with rural development required a commensurate energy policy emphasis. This was particularly crucial for those countries in which integrated rural development programmes of substantial scale and complexity were under way. The comprehensive nature of many of these programmes required substantial energy inputs,

in particular, electricity and diesel for agricultural mechanisation.

3. Based on the general policy of substituting oil by indigenous resources, rural energy development was identified with an effort to ensure adequate supplies of traditional fuels and to promote their rational use. While supply side issues concerning traditional energy concentrated on fuelwood availability, the demand side focus was on the introduction of new technologies that could make more efficient use of natural resources, such as wind, water and sunshine.
4. Decentralised renewable energy systems were widely viewed as possible alternatives to oil products to sustain the process of rural energy transition towards modern energy sources. In addition to being based on indigenous resources, these systems also carried the advantage of providing near-term solutions as opposed to the long gestation periods involved in extending centralised supplies to the rural areas.
5. Notwithstanding the emphasis of traditional fuel supplies and new energy technologies, ongoing programmes based on conventional energy systems, such as grid-based rural electrification, and the supply of kerosene and diesel, were continued to be seen as important. Thus, the development of traditional or new energy alternatives was concentrated more on those areas to which the extension of conventional energy supplies was perceived to be costly and time-consuming.

The initial set of policies (which remained in force without much change till recently) on rural energy development, therefore, adopted the strategy of continuing with conventional supply options, accompanied by a simultaneous thrust on both traditional fuels and new energy technologies. Fuelwood augmentation programmes were launched in a number of countries in which deforestation problems were more acute. These were carried out in conjunction with general forestry development programmes. Fuelwood conservation was pursued by promoting the efficiency of end-use devices mainly wood-burning cookstoves, and other applications, such as charcoal-making.

The almost universally adopted policy with regard to new energy technologies was to encourage research and development on a wide range of small-scale decentralised tehchnologies that could exploit the considerable renewable energy potential enjoyed by most countries. This policy was encouraged and supported by the international donor community. Since much of the technological research and development took place in the industrialised countries, and external funding for field-testing and demonstration was readily available from many sources, the policy to establish the feasibility of decentralised renewable energy systems was easily sustained. At the same time, the development of certain technologies, such as biogas digesters, lay within the capabilities of the developing countries themselves. Sufficient policy incentives were provided in countries like China and India to bring such technologies to a higher level of maturity.

Besides encouraging research and development activities, and bearing the full cost of equipment demonstration and pilot projects, governmental efforts were directed at disseminating technically proven alternative energy systems in the rural areas through the late 1970s and the first half of the 1980s. The policy in this regard was, however, confined to the assessment of technical feasibility rather than economic or commercial viability, since most of the diffusion took place with government or donor funding. Considerable success was achieved with biogas and small-scale hydro technologies in China, with large numbers of these systems spreading across rural areas rapidly. Other countries, especially those belonging to the Pacific island community for whom conventional supply options were limited in any case, adopted similar policies to promote technically proven systems.

At the time of the fall in oil prices in the mid-1980s, rural energy policies were still in a state of flux, with no clear orientation towards the commercialisation of successfully demonstrated decentralised renewable energy systems. On the other hand, major policy barriers, such as the stress on grid-based rural electrification and price subsidies for electricity and kerosene in the rural areas, continued. As a result, very few of the technologies involved had undergone the test of market feasibility in the commercial sense. The decline of oil prices proved to be a major setback for alternative energy technologies that had reached a higher level of maturity over the years. While policy encouragement for research, development and demonstration continued, the motivation to create the environment for a self-sustaining process of market-oriented dissemination in the rural areas went missing in most countries.

During the second half of the 1980s, rural energy development policies in many countries reverted to conventional approaches favouring grid-based rural electrification and the expansion of supply networks for kerosene and diesel. A low key momentum was maintained in the dissemination of decentralised energy systems, more due to continued donor interest and the initiative of specialised national agencies created during the high oil price period to promote these systems. However, due to their link with agricultural productivity and deforestation issues, forestry plantation programmes were pursued with vigour.

Recent Policy Initiatives

Since the beginning of the current decade, there has been a renewed emphasis on rural energy development with particular attention to fuelwood conservation and alternative energy systems. This owes to several factors, of which the following are significant:

(a) persistent gaps between the demand for and the supply of commercial energy in the rural areas, despite sustained efforts over the years to extend centralised supplies, in particular rural electrification through the grid.

(b) mounting international environmental pressures originating in concerns with the potential impacts of greenhouse gases, especially Carbon Dioxide the reduction of which is contingent to a large extent upon arresting deforestation, moderating fossil fuel use and improving energy efficiency;

(c) the refocusing of macro-economic policies in line with the sustainable development paradigm and the resultant emphasis on human development and basic needs issues;

(d) the keenness of the international donor community to integrate environmental considerations into mainstream energy development, with particular attention to the rural areas; and

(e) pressures upon multilateral development banks and other international energy financing agencies to diversify their investment portfolios incorporating energy efficiency, alternative energy development and environmental impact mitigation in a more central manner.

The above factors that have directly influenced policy changes towards rural energy development have also coincided with the economic reform initiatives of several developing countries, for example, China, India, Myanmar, Pakistan and Vietnam. The general thrust of these macro-economic reforms is to reduce the extent of governmental investment and intervention in general by encouraging the participation of the private sector in national development. (In the major planned economies, such as China, the move towards a market-based economy began in the mid-1980s and had made considerable progress by the beginning of the 1990s.

Arising from the above, a key policy shift that has taken place within the energy sector is to explore the possibilities of privatising the energy supply industry and/or inviting the participation of the private sector in energy development. Different approaches have been adopted by different countries in pursuit of such a policy shift. The participation of the private sector in developing new power generation capacity in India and Pakistan are examples of these developments.

As an extension of the policy concerning energy development in general, the involvement of the private sector in energy supplies to the rural areas has become a matter of serious interest in some countries. Although, the concept, and the mechanisms to operationalise it, are yet to crystallise, the number of private energy enterprises has been growing in some countries (India, in particular). A related development is the policy to promote self-generation and co-generation in agro-based industries (sugar mills, palm oil factories, rice mills), and to assess the potential of excess power generation that could be sold to the grid. Such arrangements have been in existence in Malaysia, Philippines and Fiji, although the electricity generated by the private sector is mainly for its own consumption and not sold to the grid. By contrast, some state power utilities in India have introduced the concept of 'energy banking' schemes to promote the development of wind energy.

Greater attention is also being paid to the participation of rural energy users in co-financing rural energy projects, in particular, projects for decentralised energy systems and fuelwood plantations.

ROLE OF FUELWOOD IN MEETING RURAL ENERGY NEEDS IN SOUTH ASIA

Above observations have indicated that biomass resources will continue to dominate the rural energy supply in South Asia.

Observations on fuelwood supply in different countries is

given below:

Bangladesh

Bangladesh has natural forest area of 769,000 ha which is about 6% of the land mass. Besides this a plantation area of 235,000 ha and other woodland of 468,000 ha also provide fuelwood. In total this is about 13.2% of the land mass. The annual rate of reforestation is reported to be 11% as against the deforestation rate of 3.3% during 1981-90. As much as 86% of all fuelwood originate from sources other than forest. In coordination with the Ministry of Agriculture, the Forest Department, under the Ministry of Environment and Forest, is also focusing on tree growing on agricultural land. Afforestation is further undertaken through the mangrow forest sources, agro forestry projects. Agro residue play a major role as a fuel source. Recently the Department of Forestry has taken keen interest in developing conversion and end-use technologies such as densification technologies for fuel briquettes and efficient wood burning stoves. The technological infrastructure, however, is reported to be weak.

Among the three sources of biomass fuel crop land and livestock are privately owned. Wood fuels are obtained from three resources grown in state owned forest land, public lands, and privately owned homestead lands. The poorer households living near state owned forests can gather fuel wood for their own consumption. Natural forests are under the management of Department of Forests. Unclassed state forests are under the ownership of the Ministry of Land and administered by the respective Hill District Councils. Village forests are privately owned and managed by the individual households.

A few constraints have been observed by the Department of Forests which include inadequate institutional capability and shortage of trained manpower, failure to design appropriate projects, insufficient project funding and inefficient forest management practices. Future fuelwood development programmes focus on reforestation, afforestation on all possible public and private lands, integrating trees with farming and traditional land use, improving the general environment for supporting agricultural and other biological production, creating employment opportunities for the landless poor, marginal savers and women and to adopt different wood conservation techniques.

Bhutan

Bhutan has natural forest area of 2,809,000 ha which is about 60% of the land mass. Besides this the plantation area is reported to be 4,000 ha and other fuelwood growing areas are reported to be 355,000 ha. The annual rate of reforestation is reported at 200 ha in the year 1990. Bhutan has a major requirement of fuelwood as 97% of its energy demand is being met by fuelwood. It has been estimated that the fuelwood requirement is at an average of one tonne per capita per year which makes the total requirement at 1 MT per year for household application. Inspite of large forest area severe localized scarcities have been reported. The capital city of Thimphu reported severe fuelwood shortage in winter in December 1995, where people had to wait up to one month for fuelwood supply. The country has no comprehensive National Energy Policy and under the old Forest Act, 1969, no popular stake is incorporated in protecting and managing forest. Lack of technological infrastructure and acute shortage of qualified manpower have been reported.

India

India has a natural forest area of 51,729,000 ha which is about 17.4 % of the total landmass. In addition a plantation area of 13,320,000 ha and other wood lands of 17,689,000 ha also contribute to additional fuelwood supply. Total amount of fuelwood reported to be consumed is 94.5 MT. Share of fuelwood from the forest varies from 26-53% and share of fuelwood from other sources is reported to 47-74%. The annual reforestation in 1990 was 1.009 mha per year. By 1993-94, deforestation has been arrested and the green cover has actually standard increasing.

In case of India, the deforestation is not due to extension of agricultural land or forest encroachment. To provide for fuelwood needs and timber needs and keeping the forest in healthy condition, there has been a deliberate Policy for importing timber from other countries; plantation of prosopis juliflora; joint forest management programme and in-farm forestry. The last two have led to a significant change in the Forest Policy and increase the fuelwood supply in the country. The Forest Policy announced in 1988, incorporates the same. Earlier, the forest lands were reserved for timber and the village for subsistence needs. Since then the forests are now designated for subsistence through joint forest management; and the farmers are encouraged to grow timber as a commercial activity.

Some areas in the country face severe fuelwood problem and the situation probably has not worsened during the last few years. In other areas, local production of fuelwood has significantly helped in overcoming fuelwood shortage. Agro residues and dung have been a significant supplementary supply sources. Integrated Energy Planning introduced under the Planning Commission has now been incorporated into the Ministry of Non-Conventional Energy Sources. MNES is also involved in developing the improved varieties of fuelwood species of tree and the technologies for end use equipment. Fuelwood based gasification and dendro thermal power generation are such significant examples. The Ministry of Environment and Forests has the responsibility for growth and maintenance of the forests and has focused its attention more on timber. A new Department of Wasteland Development is now focusing on growing fuelwood on lands that are not put to any other use.

Due to sustained efforts in these three Ministries, improvement in end use equipment efficiency, inter-fuel substitution and higher rate of urbanisation have now resulted into lowering fuelwood demand as reported by the National Council of Applied Economic Research. It is significant to note that the fuelwood demand in near future will be lower than the present demand inspite of the rural population going up in actual numbers.

India launched its planning process in 1951 in terms of Five Year Plans. Energy has been one of the major elements of public support. Different forms of commercial and non-commercial sources were effectively reflected in the plans. Share of non-commercial fuels in the total primary energy supply has declined from more than 70% in 1953 to less than 50% in 1993. The energy planning has been the major responsibility of the Perspective Division of the Planning Commission. Fuelwood has never been included in the energy planning. Integrated Energy Planning was recognised as an essential element of development planning in the 60s. The Energy Survey of India Committee was constituted in 1963 and the report was submitted in 1965. Its recommendations were considered in the Fourth Five Year Plan. The report of the Fuel Policy Committee was used for the Fifth Five Year Plan; Report of the Working Group on Energy Policy was used for the Sixth Plan and the recommendations of the Advisory Board on Energy were used for the Seventh Plan. The government has now set up an Energy Policy Committee which should be finalising the recommendations for the Ninth Five Year Plan to the period of 1997-2002. This committee is seriously considering the role of biomass fuels in the overall energy supply in the country. Primary data from the field surveys was basically carried out by the National Sample Survey Organization and the National Council of Applied Economic Research.

The domestic sector consumption worked out by ABE in 1985, indicated an energy consumption figure of 425 Kilo calories of useful energy requirement per person per day for cooking and space heating. This figure has been considered at 650 KC of useful energy requirement per person per day for cooking and space heating in the year 2004-05.

The Fuelwood Study Committee constituted by the Planning Commission in 1981 indicated the fuelwood consumption at 119 MT of oil equivalent in 1991. It is expected that this consumption will be 103 mtoe in 2004 and 84 mtoe in the year 2009. In other words the energy supply will imply continued heavy dependence on fuelwood for energy during the next 10 years. In the case of both, the crop residue and the dung, the competing demands would also increase simultaneously. It is difficult to visualise any incremental availability of either dung or crop residue without its further adverse impact on soil fertility and livestock productivity. In case the commercial energy supply fail to materialise to the extent envisaged in future, there will be much heavier dependence on traditional fuels, mainly the fuelwood.

Maldives

Maldives is a clustre of small coral islands. The total forest land and wood land is reported to be about 1,000 ha which is about 3.3% of the total land mass. It should also be noted that the total arable land is only 10% of the total land area in the year 1993. As the economy of the country is based on tourism, most of the commercial activities is dependent on the imported fossil fuel. The natives have, however, to depend on fuelwood as the major source of energy supply. The Commonwealth Secretariat has initiated action for survey of fuelwood resources and their intensification in 1987. Also under their support, designs of improved cookstoves were introduced. However, due to very little land mass for fuelwood growing, sufficient quantity of fuelwood is not available on each of the islands. Although the country has a law which makes it mandatory for anyone cutting one tree to plant two , but that is never enforced.

Nepal

Nepal has natural forest area of 5,023,000 ha which is 37% of the total land mass. In addition it also has a plantation area of 56,000 ha and other wood lands of 672,000 ha. The total quantity of fuelwood consumed is reported to be 11.3 MT per year and the share of the forest wood is 66%, whereas share of wood from other sources was reported to be 34%. The energy balance for the year 1995-96 has indicated that about 4.3 MT of fuelwood is supplied by overexploitation of the forests. The estimated area of forests and shrub land currently accessed is around 3 million ha. For the country as a whole 82% of the total fuelwood requirement is supplied from off farms and remaining 18% from the farms. In the year 1995-96, sustainable fuelwood supply from accessible forests, shrub and grass land is estimated to be about five million tonnes. 1.1 MT from farm land and 0.7 MT from non-cultivable inclusions. These supply figures indicate that around 65% of the total fuelwood requirement of the country is sustainable. Non-commercial energy accounts for 92% energy consumption in 1992-93. Fuelwood is important in Nepal for cooking as well as for space heating. The government's policy on forest management aiming to turn the forest back to village and user groups has shown successful results. New experiments have been started with regard to system of private leasehold forests.

Fuelwood requirement of urban areas has been met by forests situated up to a distance of 400 kms. The transport of fuelwood requires use of imported commercial fuel. Such transported fuelwood has more than 60% moisture content based on oven dry weight. Therefore, in the energy content scenario a lot of foreign currency has been unnecessarily drained in fuelwood supply to the urban areas located at long distances, from the source. In contrast to this in most of the rural hills fuelwood supply is non-monetized and it has been readily available in unrestricted quantities in close proximity to the consumption point.

The fuelwood development is a part of the government's policy. The last Five Year Plan (1992-97) stated that one of the main objectives of forestry sector development was "Stability of supply of timber, fuelwood, fodder and other forestry products necessary for common people in their daily life". The Forestry Sector Master Plan prepared in 1988 realised the importance of Wood Energy Plan. The document clearly identified the imbalance in the fuelwood supply and demand and suggested methods of augmenting fuelwood supplies, with an emphasis on community forestry development. Nepal has also developed a good wealth of data on fuelwood energy and rural energy supply and demand. Studies in the five different regions in Nepal on overview of fuelwood supply and demand have been completed. An Energy Pricing Study has been conducted for flow of fuelwood in Kathmandu valley. A Perspective Plan has been completed in 1995 incorporating fuelwood and biomass energy as one of the five components. Special attention has been paid to fuelwood energy planning realising that a reduction in fuelwood energy consumption is not foreseeable in the near future. The Water and Energy Commission Secretariat has been the main government department for conducting such policy guidelines.

Traditionally, fuelwood trade was carried out in many urban areas by head loaders and is in practice even today. The Fuelwood Cooperatives used to market fuelwood through its numerous depots in different urban centres until very recently. The Department of forests now controls the fuel wood cooperative. All urban fuelwood marketing and supply system in Nepal has been almost completely privatised. To some extent fuelwood supply in the fuelwood scarce rural areas is also managed by user groups under the community forest and leasehold forest programme. Under this programme, scattered patches of natural forests are being handed over to local communities for overall management by users groups with nominal technical support.

Pakistan

Pakistan has natural forest area of 1,855,000 ha which is 2.4% of the total land mass. In addition a plantation area of 168,000 ha and other wood lands of 1,105,000 ha are also contributing to fuelwood supply. The annual rate of deforestation has been reported to be 2.9% during 1981-90 and a corresponding annual rate of reforestation is reported to be 4,200 ha per year. In view of the limited forest area, the country has to focus mainly on agro forestry, which is successful now. In late 70s popular species were imported from Italy and now it is widespread in Pakistan, sustaining substantially the local wood industry and fuelwood supply. Almost 98% of all fuelwood comes from agricultural land, Agro forestry has greatly improved the fuelwood supply. About 1.5% of the agricultural land has been put under tree cover and a suggestion is made to raise it to about 5% without reduction in agricultural production. The forestry sector programme implementation takes place at the Provincial level, the Ministry of Forest and the Ministry of Environment have important role to play. The World Bank/ESMAP Study on Household Energy Survey (SHES), 1991-93 was the most comprehensive study on rural energy supply in Pakistan. Pakistan Forestry Institute has a strong technology development and technical manpower base.

Sri Lanka

Sri Lanka, has natural forest area of 1,746,000 ha, which is about 27% of the land mass. Besides this, a plantation area of 139,000 ha and other wood lands of 2,113,000 ha also contribute to additional fuelwood supply. The biomass fuel consumption contributes to 53% of total energy consumption in 1992. The annual rate of deforestation is reported to be 1.3% during 1981-1990. The country has national forestry programme which also address bio-energy as an important energy source for Sri Lanka. Biomass including fuelwood and charcoal, figure in the national energy balance of Sri Lanka for the last 10 years. This has been confirmed by the Ministry of Irrigation, Power and Energy. An important programme of improved woodstoves was introduced which was supervised and implemented by the Electricity Board. The stove programme for both the rural and the urban areas is now promoted with the help of private entrepreneurs.

ROLE OF BIOMASS—CERTAIN FACTS

Prof. W.S. Hulscher, CTA, RWEDP in his presentation on "Sustainable Energy Supply Options", delivered at the Asia Energy Vision 2020 at New Delhi during November 15-17, 1996 stated as follows:

The cheapest and most reliable option for supplying renewable energy is still to grow a tree and burn it. This is actually what is commonly done in Asia, even though tree growing in seldom meant for the sole purpose of supplying energy. Unfortunately, the important roles and associated problems of wood and other biomass as a sustainable energy supply option are largely undervalued by energy planners and policy makers. A few basic data will suffice to substantiate the point. Table 1.1 overviews the share of wood and biomass energy consumption in 7 countries in South Asia. Table 1.2 shows the trend in biomass energy consumption. In Table 1.3 the total value of woodfuel consumption per annum is estimated in terms of dollar equivalent. Table 1.4 summarises some figures on estimated employment per fuel type. These data lead to the following important observations:

(a) Wood and other biomass suply about one-third of the national energy consumption in India, and some 37%

of the energy consumption in the region (in South and South-East Asia). Further comparisons show that amongst the various renewable energies wood/biomass provides by far the largest share in the region.

TABLE 1.1: **Share of Biomass and Wood Energy Consumption**

Country	Year	Share of Biomass (%)	Share of Wood (%)	Share of Biomass in Domestic Sector (%)
Bangladesh	1992	72	13	89
Bhutan	1991	82	—	—
India	1992	33	—	78
Maldives	—	—	—	—
Nepal	1992-93	92	68	97
Pakistan	1993-94	47	27	83
Sri Lanka	1990	77	—	93

TABLE 1.2: **Biomass Energy Consumption (PJ) in Member Countries**

Country	1981	1986	1991
Bangladesh	243	162	277
Bhutan	7	9	12
India	2,165	2,441	2,824
Nepal	113	197	206
Pakistan	192	233	296
Sri Lanka	70	78	86
Maldives	—	—	—

TABLE 1.3: **Woodfuel Values in Million US$ using Average Woodfuel Prices**

Country	Fuel-wood	Char-coal	Dom. FW	Dom. Char	Ind. FW	Ind. Char	Total
Bangladesh	306	—	255	—	50	—	306
Bhutan	37	3	33	—	3	3	40
India	9,080	—	8,440	—	640	—	9,080
Maldives	3	—	3	—	—	—	3
Nepal	469	—	451	—	17	—	469
Pakistan	1,318	—	1,317	—	—	—	1,318
Sri Lanka	436	5	363	—	—	—	440

TABLE 1.4: **Estimated Employment Per Fuel Type**

Fuel type	Tons of Fuel per Tera Joule (T)		Estimated Employment per TJ Energy Consumed in Person Days'
Kerosene	29 Kilo	Litre	10
LPG	22	Tons	10-20
Coal	43	Tons	20-40
Electricity	228	MWh	80-110
Fuelwood	62	Tons	110-170
Charcoal	33	Tons	200-350

(b) In all the named countries the consumption of wood and other biomass fuels is still increasing in absolute terms, even when their share in national energy consumption is decreasing.

(c) The total value of wood and biomass fuels amounts to

more than 9 billion US $ per annum in India (which was 30-40% of export earnings in 1991), and as much as some 12 billion US $ per annum for the named countries in the region.

(d) Woodfuel use generates at least 20 times more local employment than energy from oil products (per unit of energy). Hence, national interests like balanced growth, employment and equity are linked to energy supply policies.

ENVIRONMENTAL CONCERNS

Energy conversion has a significant impact on health and environment. As the rural energy consumption in South Asia is mostly in terms of cooking energy, the impact of burning coal and biomass indoors has been studied exhaustively. These studies have been done on the basis of the assumption that most of the cooking is done indoors. Due to tropical climate this is not wholly true in the South Asia region. UNDP in its report "Energy After RIO" (1997) has focused that the energy-health nexus arises from the fact that, without proper control, the production and use of energy can be accompanied by adverse impacts on the environment and ultimately on human health. The amount and type of energy used is itself general indicator of the potential for a number of important environmental health risks.

Fundamental to an evaluation of the environmental and health impact of energy is the concept of fuel (or flow) cycle; i.e., the chain of activities that extend from energy source extraction, through processing, transport and storage to the NGOs.

In cases such as the gathering of agricultural waste for fuel in a Bangladesh village, the spatial domain and time period over which the fuel cycle operates is short. What is gathered today is burnt nearby tomorrow. At the other end of the spectrum, however, are the large scale fuel cycles for conventional energy which link activities occurring over many parts of the globe, extending over months, years, or even centuries in the case of waste disposal.

Human impact on the global environment mainly concerning the portion attributable to energy supply can be summed up as follows:

Sulphur dioxide flow contributes to 1 per cent due to burning of additional fuels, methane stock contributes 5 per cent burning of traditional fuels. Nitrous oxide contributes to 8 per cent burning of additional fuels; particle flow to 10 per cent burning of traditional fuels and carbon dioxide flow to 3 per cent of the net deforestation for fuel wood. UNDP has also worked out the human destruction index which is defined of the ratio of human generated flow to the natural baseline flow. This is the highest in case of sulphur dioxide flow and is the lowest in case of carbon dioxide flow. Basically the human destruction due to burning of the traditional fuels is at a low level due to larger use of traditional fuels. However, it has to be noted that the nominal dose effectiveness due to mainstream cigarette smoking is the highest. The next to it is stove vented indoors and subsequently the stove vented outdoors.

In its study on household use of biomass and coal the UNDP has estimated that biomass accounts for 1/3rd of all energy and nearly 90 per cent in some of the least developed countries. Out of 1.25 billion people living in South Asia about 900 million are reported to be in the rural areas and almost about 1 billion are still dependent on biomass fuels for their daily energy needs This includes people in the rural and semi urban areas and the urban slums. In the rural areas the cooking is done generally in the open or a well ventilated place. But in the semi urban areas and the urban slums, cooking is normally done in an enclosed place either due to the urban designs of the kitchens or due to lack of space. The stove is often at floor level, adding to the risk of accidents and jeopardising food hygiene. In many instances, there is no chimney to remove the pollutants.

Solid fuels such as biomass are quite difficult to burn completely in simple household sized stoves as a result, even though biomass does not contain many non-combustible contaminants, the emissions of health damaging pollutants in the form of incumbustible products are quite high per unit of energy. The use of fossil fuels for space and water heating in the home, or for cooking, is also widespread in many countries.

Typically emissions are high in the rural households in South Asia. Because occupants usually are indoors when they are using coal and biomass, they tend to be exposed to significant amounts of particulate pollution. Because a large portion of the population is exposed, the total indoor air pollution exposure is likely to be greater for most important pollutants than from outdoor urban pollution in all the world cities combined.

Most of the remote villages will be connected through metal roads and if the agro processing and agro based industries expand as planned, the road traffic will substantially increase increasing the risk of pollutants due to fossil fuels burning in the transport sector.

Health effects in females have been largely observed in certain surveys carried out in the rural areas in India. The largest impacts have been reported in terms of respiratory infections in children and chronic lung disease in non-smoking women. There are also a range of other health problems associated with this fuel cycle, some direct such as the impacts on women and children of gathering heavy loads of biomass in distant and sometimes dangerous areas.

In addition, indirect health impacts from lack of fuel for proper cooking (malnutrition) and boiling water (diarrhoea and parasites) may be significant, although difficult to document.

Fortunately, the use of coal is on a limited scale and the problems due to coal burning are not acute. It has been suggested that the conversion processes should be designed that maximum efficiency is achieved leading to lower amounts of pollutants.

Energy policies in South Asia have traditionally been based on making use of indigenous energy resources to the maximum extent possible. In many parts of the region, it has been observed that the recent economic development has led to emergence of semi-urban, fully developed and self-dependent areas. There is no marked distinction between the urban and rural areas as a process of continuum is observed. This, therefore, requires that as far as pollution issues are concerned for the rural energy supply the impact of development of urban centres cannot be ignored. Most important of the activities is the transportation sector, which is the basic link between the urban and the rural areas with emphasis of infrastructure building for economic development of rural areas and new thrust is being placed on the transportation sector.

The countries of South have the natural and human resources, including well trained manpower, necessary to meet the twin goals of energy supply and environmental quality.

Another aspect that needs to be considered is that the inefficient burning of biomass emits smoke causing health hazard to beneficiaries who are largely in the lower income groups of the rural and urban areas. In the normal course, the major objective of energy planning should be either to replace these traditional fuels completely or introduce cleaner and more efficient technologies for the conversion of biomass into useful energy for cooking. The latter option will call for major investments as well as technological upgradation. In the case of dung, biogas is a well established technology. If the traditional fuels are to be replaced completely, it will require either kerosene or coal-based fuels. In other words, to protect the rural and the urban poor from exposure to excessive emission of smoke, it is necessary to step up fossil fuel supplies, either petroleum products or coal or both.

Even from the point of view of the long term effect of carbon dioxide emission on the environment, it should be noted that one tonne of coal used in the household sector will save 2.25 coal-equivalent tonnes of fuelwood as the end-use efficiencies of the two fuels are in the ratio of 2,25:1. Similarly, one tonne of kerosene used in this sector will save 5.6 kerosene equivalent tonnes of fuelwood. This is a typical example of how fossil fuel consumption at the margin can be environmentally benign in developing countries like India.

While such transition from traditional to commercial fuels will, therefore, be welcome from the point of view of local as well as global environmental concerns, it is beset with other problems such as affordability of the low income groups who will then be required to use the more costly commercial fuels in the place of non-commercial fuels. According to one study, the expenditure of household energy consumption in the lower income groups in both rural and urban areas amounts to 10 to 14% of the total household expenditure on all items. Increased dependence on commercial energy will, therefore, impose a heavier burden on the poor. The only way to resolve this problem is to enhance the purchasing power of the low income groups of the community.

Air pollution in many of the largest cities in South Asia greatly exceeds the level considered acceptable by the World Health Organization. Calcutta, Delhi, Karachi, Kathmandu, Lahore and Mumbai are among the cities with the highest levels of particulates or other air pollutants. In addition to these local pollutants, the contribution of the region to the emission of green house gases, particularly Carbon Dioxide is also rising. Asian emissions of Carbon Dioxide now exceeds those of North America, Western Europe or Eastern Europe including the Russian Federation. On a per capita basis, of course, Carbon Dioxide emission from South Asia are only a small fraction of those from the industrialised countries, and are likely to keep increasing, well into the 21st Century. However, all the South Asian countries have signed the framework convention on climate change. The convention does not require countries to set limits on emissions of green house gases, but signatories from the developing countries are expected to identify policy options that could reduce future emission of green house gases from the region.

INTERVENTION AND INSTITUTIONS

Interventions in the wood energy sub-sector largely depends on government actions like public resource management, pricing, taxes, subsidies, economic incentives etc. Such interventions can only be prepared and implemented when the country avails of adequate institutions and sufficient number of qualified experts. Further more interventions, with regard to wood energy are to be supported by the establishment of effective public-private partnership for implementing new supplies and managing existing ones, as well as private sector organisations for production, conversion, utilisation technologies.

With most of the countries in South Asia opening up their economy and promoting larger private sector participation, public policy interventions have become all the more important. The initial phase of energy sector privatization has brought forth a number of complicated issues starting from legal contracting to power purchase agreement, counter guarantees, sustained fuel supply etc. It has a direct relevance to the rural energy sector. The government's emphasis and future extension in all the countries is for electrification or energisation of the left out areas of which most at the rural areas in the South Asian countries.

The approach also has a direct bearing on promoting the fossil sources of energy or the renewable source of energy. Even amongst the renewable sources, the place of wood energy, which is of direct relevance to rural areas has still not found appropriate place. More so fuelwood does not find an appropriate place either amongst the ministries dealing with energy or those dealing with agriculture and forestry.

Next comes the development of skills for units dealing with developing, planning and managing wood and biomass energy. In order to gain momentum a critical mass of educated and well trained people will be necessary. Barring India, no other country in South Asia has skills for all forms of fossil and renewable forms of energy. Thus there is a need for developing appropriate energy education curricula.

Further important aspect of institutional development relate to the rolls of State and local Government authorities, NGOs, Community participation and private sector organizations.

CHALLENGES TO WOOD ENERGY PLANNING

As a part of RWEDP strategy inference, Prof. W.S. Hulscher has specified following challenges:

Wood energy qualifies as a modern option when it is efficient, clean, convenient, reliable and economically and environmentally sound. Many of the widespread practices associated with fuelwood energy are still far from these qualifications. This poses many challenges to the ingenuity of not only scientists and engineers, but also specialists in natural resource management, forestry, agriculture, economy, planning, environmental management, social sciences, extension etc, as well as the integration of the elements of the named disciplines. For instance, the use of fuelwood and biomass in rural, small and medium scale industry is widespread in South Asia. These enterprises are vital for the economy in terms of product output, employment, rural sustenance and entrepreneurship. However, fuel used in these units suffers from many problems, like inefficient combustion, unhealthy working conditions, unreliable supplies, poor financial viability, lack of technical and managerial skills etc. The response to these challenges lies in identifying alternative forms of energy. This type of work requires integrated efforts and skills of many of the above named specialists.

The wood energy planning to be feasible at all has to be based on decentralized structures for example linked to area based planning or to the rural development planning. Decentralization allows close interaction between planning and implementation, whether in projects, programmes and policies at large. Experience has shown that such interaction is viable for successful interventions which address local and site specific issues like wood energy. Wood Energy Planning has both long term and short term components that is from one to ten years or more.

For fuelwood planning, systematic data is required as a time series, which implies more than some casual survey. Data availability is a problem in most countries. The relevant data are scattered over various institutions and they are not as accurate as the data readily in hand for fossil fuel. Data are sufficiently accurate if atleast they allow a correct understanding of problem at stake and provide grasp of trends over time. These two requirements can be met by a combination of micro studies at certain intervals and well selected micro studies. Relevant micro studies would be on wood energy consumption patterns and trends, (in households, industries, institutions and commerce etc.), and on wood energy supply patterns and trends, including studies of wood fuel sources, flows and markets.

As usual in the energy sector, planning is taken as matching supply and demand. On the supply side wood energy planning calls for quantitative data on supply patterns and trends of wood and other biomass fuels. Such data include the sources of supply, their productivity and sustainability, and the patterns of woodfuel flows from sources to markets and finally to consumers. Related socio-economic data are on employment (or income generation) and gender aspects. Linkages with geo-information on land use from forestry, agriculture and environment sectors backs the interpretation of data obtained from supply surveys. Part of the data can be obtained from aerial surveys, but many aspects also need ground surveys. The data are to be compiled in data bases providing overviews for planning and policy making.

On the demand side wood energy planning requires quantitative data on consumption patterns and trends of wood and other biomass fuels in a systematic way. Such data, again, can only be obtained by surveys. The data need to be linked to information on the local rural and national economy, and key information from relevant sectors like energy, forestry, agriculture, and environment. Relevant social conditions, gender related information, and distributional aspects must be studied to help interpret consumption data. The surveys should lead to quantitative overviews to be installed in data bases which are accessible for purposes of planning and policy making.

Incentives to Promote Alternative Energy Systems

In line with the foregoing policy changes, a new range of fiscal and financial incentives to promote alternative energy systems is in the process of being evolved in the countries surveyed. While some of these incentives are in the form of direct governmental support, as in the past, several are recent innovations. A summary of currently available incentives for alternative energy systems is provided in Table 1.4. Based on this, the following observations are made on the prevailing policy climate towards alternative energy development:

— Significant policy change across the spectrum of decentralised renewable energy technologies has taken place only in (China and) India. Both these countries have a longer history of technology diffusion and familiarity with a wide variety of technologies.

— Other countries have taken more cautious steps, offering encouragement on a selective basis to one or just a few proven technologies. Bangladesh and Nepal have sought to encourage solar photovoltaic systems, while Myanmar, Thailand and Vietnam have favoured small-scale hydro systems. Some of these countries have also provided inducements for biogas (Nepal, Myanmar and Vietnam), while Thailand is the only one to promote wind water-pumping systems. On the appliance side, improved cookstoves have received the most frequent attention.

— The foreign origins of some technologies, especially solar photovoltaic systems and system components, has been recognised and reflected in customs duty and tax rebates in all countries that have favoured this technology.

— There is a clear policy shift in several countries to move away from the erstwhile government-funded demonstration project syndrome to a new paradigm of private sector-oriented and end-user-financed market diffusion. Concessional loans have been introduced in a number of countries and an attempt to cut back on governmental investment in well-developed technologies has been made in China.

Future Thrusts

The main conclusion of the foregoing review is that there are clear indications of change towards a more favourable policy environment for the accelerated development of alternative rural energy systems. However, this is offset by the perpetuation of policy barriers, in particular, capital cost and pricing subsidies for conventional energy supply systems , and the scarcity of front-end capital for investment in alternative energy systems. The main thrust of future rural energy development policy would be to remove or reduce the effect of such barriers to facilitate a greater role for decentralised renewable energy technologies.

Subject to the distinctions among individual countries, a range of policy measures would be needed on a regional basis to promote decentralised renewable energy systems under future rural energy development efforts. While these measures would have to be adopted and pursued by the countries themselves, they would require external encouragement and assistance from the international donor community and multilateral development banks because of the capital and foreign currency intensity of most of the technologies involved. The following is an indicative list of such measures:

1. The volume of governmental investment in rural energy development as a whole may be increased, with specific emphasis on areas that are unlikely to attract private sector investment (for example, fuelwood plantation, improved cookstoves) or where governmental investment would help create the core technology markets and act as an inducement to the private sector for down-stream market development and expansion activities.
2. In areas where private sector investment interest is evident, a positive climate for investment could be created through policy incentives such as pioneer status, depreciation allowance and area franchises. Joint ventures involving foreign renewable energy technology manufacturers may be encouraged through liberal equity participation and capital/profit repatriation facilities. Completely indigenous ventures could be allowed the necessary foreign currency facilities to acquire alternative energy systems, and the capital equipment and machinery needed to manufacture such systems, or parts thereof, locally.
3. Licensing requirements for the local manufacture and assembling of renewable energy systems and system components may need to be eliminated or liberalised. This should be accompanied by arrangements for the expeditious clearance of site permits and other

procedural requirements.

4. National development banks may be required to take the lead in diversifying their credit portfolios to finance decentralized renewable energy systems. Alternatively, governments may set up financing institutions specialising in credit for rural energy development. These agencies would have to acquire the capabilities to act as retail on lending intermediaries for funds provided by multilateral development banks and international donor sources. The financing arrangements should address the needs of private sector equipment manufacturers, importers and dealers, with a special emphasis on end-user financing.
5. Governmental investment in state electricity utilities may be regulated in proportion to the planned mix between centralised electrification and electrification based on decentralised options. In other words, substitution effects may be more clearly reflected in the investment allocation for utilities in order to ensure that decentralised rural electrification is not seen as a marginal addition to grid-based rural electrification.
6. The existing subsidies in the prices of commercial energy sources used in the rural areas—in particular, the prices of electricity, kerosene and diesel—may either be phased out or significantly reduced in order to reflect the long range marginal costs of supply. However, where the elimination of such subsidies in the short term is not found to be politically and/or socially viable, countervailing subsidies may be extended to technically proven and economically viable decentralized renewable energy technologies. Such subsidies could be restricted to only those technologies for which financial or commercial viability is the principal constraint because of subsidies inherent in the prices of competing centralized sources of supply. Since the main end-user

cost in decentralised renewable energy systems is at the initial investment stage rather than at the stage of system operation, countervailing subsidies should take the form of capital cost rather than pricing subsidies.

7. More liberal fiscal incentives may be provided for imported technologies in the form of waived or reduced customs duties and other import taxes. These incentives would apply to a wide range of proven technologies that are considered suitable to meet the rural energy needs of individual countries and should cover complete systems, completely knocked down systems for local assembly, specific components (for example, solar cells and battery control units) as well as spare parts for maintenance in order to encourage the share of local value added and to promote local manufacturing capabilities, the incentives provided may be graduated according to their local capital cost component on a variable basis.
8. In view of the foreign origin of many renewable energy technologies, the policy focus on technology transfer may be sharpened. This would require close co-operation between the developing countries and the industrialised nations in which such technologies originate. Policies in this regard may emphasise the rapid development of local manufacturing capabilities for which, among other measures, a well-supported application-oriented research and development base would be essential.

ENERGY FROM BIOMASS: PRESENT STATUS IN NEPAL

The total energy consumption in Nepal in the year 1995-96 was about 292 million GJ, of which about 90% was supplied from biomass sources, such as fuelwood, agricultural residue and animal dung. Fuelwood among the biomass energy sources, currently represents about 89% of the total consumption. In the fuelwood scarce areas, agricultural residue and animal dung also represent considerable share in rural energy consumption.

A study "Rural Household Energy Use" was conducted by the Water and Electricity Commission Secretariat of Nepal in 1986, covering the whole terai region and one hilly region of Mid Western Development Regions of Nepal. WECs compile energy consumption data for Nepal based on this study with the help of supplementary information from Community Forestry Development Project and Australian Forestry Project etc. The share of fuelwood, agricultural residue and animal dung in the total energy consumption was estimated and has been in use for general planning in Nepal. (Energy Sector Synopsis Report, 1992-93, WECs). Being the only information available, comprehensively it served as base line data for general energy planning in Nepal. The share of fuelwood, as per the new study, which is based upon comprehensive number of samples, compared to previous information source has been estimated, higher by about 10%. Similarly, share of agricultural residue and animal dung have also changed dramatically. The total share of biomass energy has decreased by about 2%. Major explanatory factor in urbanisation rate, which is much higher than population growth.

Reasons for change in share of biomass energy sources: The share of fuelwood has increased during the period, while the area under forest resources, shrub lands, non-cultivate inclusion (NCI) etc. are being reduced rapidly with the increasing population and economic activities. The previous information, "Rural Energy Household Survey, 1986" was based on a few hundred samples and covered on the Terai region and one hill region of MWDR. The latest information compiled by WECs is derived from comprehensive number of samples, that exceeds 12,000. A careful scrutiny will reveal that both the agricultural residues and the animal dung are commonly used in the Terai region and its use in the hills and mountain region is nominal. But, lack of information on hills and mountains, energy consumption has possibly caused a different projection in the past. Similarly, difference in shares of agricultural residues and animal dung compared to previous information is also alarming. Reasons assigned for such deviations in findings are: (a) lesser coverage, (b) availability of fragmented information.

BIOENERGY CONSUMPTION IN DIFFERENT SECTORS OF THE ECONOMY

The overall energy consumption has been projected in residential, industrial, commercial, transportation and agriculture sector. The share of biomass in the total energy consumption in different sectors also provide an insight into linkage between Energy and Economy. Residential sector is the largest energy consuming sector, that account for about 90% of the total consumption, in which the biomass contributes about 98%. Industrial and commercial sectors are heavily dependent on biomass resources for energy requirement. Nearly 30% energy requirement of these sectors come from biomass. There is no contribution of biomass in transport and agriculture sector.

Share of residential fuelwood use in urban and rural areas are about 75% and 98% respectively. Consumption of biomass sources in residential sector for energy purpose depends heavily on its availability, price and convenience in use. In urban areas too, the fuelwood use varies widely. For example, fuelwood used in Kathmandu is only about 2%, whereas fuelwood used in Bharatpur is about 76%. The availability and price of the fuelwood, level of economic development and also convenience factors affect the level of fuelwood use.

A substantial portion of biomass sources are still being used in the industrial sector for energy. About 30% of the energy in this sector is in the form of fuelwood and agricultural residues. This indicates low upgradation or modernisation of this sector, and switching over to other modern fuels. Biomass is basically used in industrial sector for the process heat and boiler. The other biomass consuming sector is the consuming sector, which includes restaurants, institutions, hotels, hospitals, cinema hall, water supply services, essential and non-essential shops etc. The energy consuming of this sector are located mostly in the urban areas, which are mainly service oriented and account for about 31% of its energy requirement.

END USES OF BIOMASS ENERGY

The biomass energy is used mainly for cooking purpose in the domestic and commercial sectors. The next important use is water heating (boiling) and space heating. Biomass energy is also used for agro-processing, animal feed cooking, and rituals in the residential sector. The percentage share of biomass energy in different end uses is as follows:

Except for some end uses of industrial sectors (that is for boiler and process heat), biomass energy use exceeds 90%. Energy required for agro-processing and animal feed cooking are met almost entirely by the biomass sources. Similarly the biomass energy share for cooking and heating are 97 and 100% respectively. In the industrial sector, biomass energy for the boiler and process heat applications account for 42 and 63% respectively.

It has also been observed that most of the biomass is burnt in the traditional wood stove, specially in the rural areas. It is observed that the traditional stoves have multiple uses in terms of cooking food, cooking animal food, space heating, water boiling etc. Though different programmes of dissemination of improved cooked stoves were launched in the past with the objective of reducing fuelwood consumption through increased end use efficiency, the impact of these programmes has almost negligible. It is also observed that these stoves are not widely used for a number of social and other reasons. Inconvenience in use of improved stoves, for multiple applications is the most important reason. Also lack of R&D facilities in the institution for appropriately adopting the cook stoves to the site condition is another important reason. People, however, will always like efficient and modern devices, if it serves for all uses that it intends to substitute.

Some of the energy end-uses, where biomass energy is not found suitable, include, cooling, lighting and electric appliances. unlike residential sectors, many devices for cooking, water boiling, heating in the commercial sector are non-biomass base as they are more convenient to use and being in the monetized sector, the user can afford other convenient form of energy.

Biomass Energy by Development Region

Mid Western Development Region (MWDR) is the biggest one and smallest is the Far Western Development Region in size. But, CDR consumes largest share of total energy followed by the WDR, EDR, MWDR and FWDR. Looking at the biomass side, FWDR meets about 97% of its energy requirement from biomass. FWDR is followed by MWDR, WDR, EDR and CDR in biomass dependency. This is because of the higher affordability of the people and economy to substitute biomass energy by more modern forms of energy. In vice versa, higher share of commercial energy in overall energy consumption also indicates higher level of access to commercial energy and economic activities. However, access and availability of the biomass resource, mainly forest, also has bearing in the share of biomass energy in overall energy consumption. This can be substantiated from the fact that the per capita forest area is less in the CDR than other developed regions. Another factor that affects this share is the urbanization rate, which is the lowest in FWDR and the highest in CDR.

Energy Consumption in Different Physiographic Region

Nepal is divided broadly into three physiographic regions—southern plain areas (Terai), Hills in the middle and

Himalayas or Mountains in the north. Terai is economically important and active compared to Hills and Mountains as the region has better facilities in terms of power, road and access to market, etc. Obviously, it affects on the total energy consumption pattern. It is observed that the total energy consumption is very high in the Terai region compared to other followed by Hills and Mountains. But, it will be contrasting if we observed the status of biomass energy in each region.

NOTES

Discussion with the Engineers from the Bangladesh Institution of Engineers.

Discussion with the Officials from the Government of Bhutan.

Discussion with the Secretary, Pakistan National Committee, World Energy Council.

Discussion with the Secretary, Srilanka National Committee, World Energy Council.

Energy after Rio: Prospects and Challenges, UNDP, January 1997.

Human Development in South Asia 1997, Mahbub ul Haq, The Human Development Centre, February 1997.

National Study Plan on Integrated Rural Energy Planning and Environmental Assessment (for Srilanka), Draft Report Presented to the National Orientation Workshop at Colombo in December 1997, UNDP/ESCAP PACE-E REED 2.

Periodic Journal of the Water & Energy Commission Secretariate, Government of Nepal.

Projects and Programmes 1998, Ministry of Irrigation and Power, Government of Srilanka.

Regional Study on Wood Energy: Today and Tomorrow, RWEDP-FAO Field Document No. 50, October 1997.

Rural Energy in the Asia-Pacific Region, RAP Bulletin, 1997-98.

Rural Energy Systems in the Asia-Pacific—A Survey of their Status, Planning and Management, Ed. K.V. Ramani *et al.*, APDC, 1993.

The State of Forest Report 1993, Forest Survey of India, GOI.

The Status of Watershed Management in Asia, FARM-UNDP/FAO, August 1995.

The Woodfuel Scenario and Policy Issues in India, N.C. Saxena, RWEDP-FAO-Field Document No. 49.

2

WOOD ENERGY SYSTEMS IN INDIA

SECTION I

CURRENT SITUATION OF FUEL WOOD PRODUCTION AND USE IN THE COUNTRY

1.1 Wood Resource Supplies

The national forest policy, 1988 was announced in December 1988 with the principle aim of ensuring environmental stability and maintenance of ecological balance. The Forest Conservation Act, 1980 was amended in 1988 to facilitate stricter implementation and to plug certain loopholes. The rate of diversion of forests plan was brought down to about 0.017 million hectares per year from 0.15 million hectares per year during the period 1951-52 and 1979-80.

The state of the Forest Report 1993 projected the assessment of the forest cover of India based on visual and digital interpretation of the satellite data pertaining to the period 1989-91. According to this assessment, the total forest cover of the country is 6,40,107 sq. km (excluding tea gardens) which is 19.47% of the total geographical area of the country.

The mix of wood resources by supply source are given in the Table 2.1.

1.2 Role of Energy in Overall Wood Resource

To have an insight into the production of wood energy

Paper submitted to the Forum "Forest and Energy" organised by Welt Forum Wald held at Schneverdingen, Germany during January 17-20, 1998.

TABLE 2.1: **Mix of wood resources by supply source - Production (1993) mt/yr.**

	Production	Average Annual Growth Rate (1983-93) in %
Roundwood	28.75	1.9
Fuelwood & Charcoal	262.78	2.0
Industrial Roundwood	24.67	0.8
Sowing & Veneer Logs	18.35	0.6
Sownwood & Sleepers	17.46	1.2
Wood Based Panels	0.44	-0.6
Wood Pulp	1.15	3.6
Paper and Paper Board	2.63	5.1

Ref: Selected Indications of Food and Agriculture Development in Asia-Pacific—1984-94 UNFAO Regional Office, Bangkok—1995.

resources, it is also necessary to have an idea about the overall energy consumption and then to know the contribution of wood and biomass energy balance. The energy balance in India in Peta Jules in the year 1993-94 is shown below.

Overall wood resource supply is given in the Table 2.2.

TABLE 2.2: **Consumption of conventional, wood and biomass energy in 1993-94**

UNIT: PJ	Total Energy	Conventional Energy	Woodfuels	Biomass Energy	Share of woodfuels in Total	Share of Biomass in Total
India	8,751	5,822	2,603	2,929	30%	33%

Source: Estimated from data of IEA, WRI, Country data and RWEDP's Best Estimates.

The consumption of biomass fuels is increasing, whereas the share of biomass energy in the total energy consumption is declining in India. The two major factors influencing the consumption pattern are:

(i) liberalisation of markets has been more dispersed industrial activity, requiring higher energy consumption from commercial sources of energy.

(ii) the increasing government support price for the cereal food productions in the rural areas and affluent class is opting for convenient cooking fuels like LPG and Kerosene.

The basic observation is the combination of interfuel substitution for a small extent for the prosperous farmers; and supplementary energy supply for off-farm income generation activities through the commercial sources of energy.

The energy balance does not clearly distinguished between rural and urban households and energy supply to large scale and small and tiny industries. Such a distinction would be relevant for wood and biomass energy, since rural households and small scale industries are generally the main wood and biomass energy consumers.

1.2.1 Availability of Wood Resources by Main Agro-Ecological Zones

The only documented information on this aspect is from the report of the Advisory Board of Energy, Government of India, published in 1984 with the title "Towards the Prospective of Energy Demand and Supply in India in 2004-05". Additionally the Rural Energy Data Base made an estimate of biofuel consumption in different agro-climatic regions under the IREP programme. This assessment was made in 1992. The production figures are expected to match with the consumption figures. This is the only authentic summary of biofuel consumption produced by the Planning Commission. The details are given in Table 2.3.

The National Council of Applied Economic Research also made a statewise biofuel consumption projection in the year 1985. The same are given in the Table 2.4.

TABLE 2.3: **Summary of biofuel consumption in agroclimatic regions (IREP estimates)**

Agroclimatic Sone	Firewood	Crop residue	Dung cake
Western Himalayan Zone	1.95	0.14	0.22
Eastern Himalayan Zone	1.70	0.00	0.00
Middle Gangetic Zone	0.75	0.65	0.50
Upper Gangetic Zone	0.36	0.35	0.43
Trans Gangetic Plains	0.37	0.96	0.65
Eastern Plateau and hills	1.60	0.00	0.00
Western Plateau and hills	0.79	0.38	0.47
Southern plateau and hills	0.62	0.30	0.19
East coast plains and hills	0.60	0.08	0.13
West coast plains and ghats	1.29	0.21	0.00
Gujarat plains and hills	0.92	0.30	0.22
Western dry regions	1.08	0.03	0.44

Ref: Review of IREP - Planning Commission 1992

TABLE 2.4: **State-wise biofuel consumption (Kg./Capita/Day)**

State	Firewood	Dung cake	Agricultural residues
Andhra Pradesh	0.55	0.13	0.10
Arunachal Pradesh	0.31	0.00	0.20
Assam	0.31	0.00	0.20
Bihar	0.33	0.73	0.22
Goa	0.31	0.25	0.04
Gujarat	0.40	0.25	0.02
Haryana	0.14	0.70	0.21
Himachal Pradesh	1.31	0.02	0.00
Karnataka	0.60	0.02	0.14
Kerala	0.41	0.00	0.23

(Contd.)

TABLE 2.4 *(Contd.)*

State	Firewood	Dung cake	Agricultural residues
Madhya Pradesh	0.47	0.52	0.12
Maharashtra	0.31	0.25	0.04
Manipur	0.77	0.00	0.23
Meghalaya	0.77	0.00	0.23
Mizoram	0.77	0.00	0.23
Nagaland	0.77	0.00	0.23
Orissa	0.46	0.33	0.08
Punjab	0.25	0.58	0.32
Rajasthan	0.57	0.39	0.04
Sikkim	0.31	0.00	0.20
Tamil Nadu	0.47	0.11	0.11
Tripura	0.77	0.00	0.23
Uttar Pradesh	0.38	0.58	0.17
West Bengal	0.25	0.27	0.43
UNION TERRITORIES	0.25	0.36	0.17

Ref: NCAER 1985

TABLE 2.5: **Indicative sources of fuelwood household (HH) and industrial (Ind.) use as % of total amount used in 1996**

Country	Year & Sector	Million Tons	Forest land	Other land	Public land	Unknown
India	1996, HH	162	51	49	—	—

Source: Ministry of Environment & Forests, 1996.

Since oil crisis took place, there has been a continuous effort at the international level to indicate that deforestation is taking place at a very fast pace, mainly due to higher consumption of fuel wood for cooking in the developing countries. The 'Gap Theory' was so strongly accepted that by the time 1981, Nairobi Conference took place, the major agenda was focused on deforestation due to fuelwood consumption. The 'Gap Theory', took strong roots and was often quoted in the past to justify actions in the field of enhancement of forest resources. Fuelwood energy conservation programme was based on the belief that most, if not all, wood fuel originated from forest. Fortunately at this point of time the world has a data for almost 15 years of indepth studies that have shown that non-forest areas supply considerable amount of wood fuels. In fact, RWEDP has collected evidence, "albeit sketchy" shows that in many countries, a major part up to 50% of wood fuel is derived from non-forest areas. The latter include barren land, agricultural land, agricultural crop plantation (Rubber, Coconut etc.) homesteads, trees along road sides etc.

1.2.2 Availability of fuelwood by type of wood management system

Fuelwood is generally gathered by rural people on both public and private lands. The share of each source of fuelwood supply has been developed by NCAER as follows:

A recent survey for the primarily sources of fuelwood shows that the total consumption of fuelwood in 1988-91 was 172.5 MT. As indicated in the above two tables, a large portion of the fuelwood continues to be, by and large, a non-monetised commodity. Even when firewood is traded the study showed that rural fuelwood markets are small and localised and lack of capital and buying capacity. The fuelwood supply to the market is only about 25% of the consumption. There are two implications for the production of fuelwood as a farm crop.

1. First, gatherers can always undercut producers in the pricing of fuelwood; the producers would be price takers, rather than price makers.
2. Second, market prices of fuelwood would always be lower than its social cost for replacement of growing stock through investments in plantations.

TABLE 2.6: **Sources of supply of firewood in India in 1978-79 in MT per year**

	Rural			Urban			Grand Total
	Log	Twigs	Total	Logs	Twigs	Total	
Collected from							
i. Own land	5.2	9.1	14.3				14.3
ii. Neighbour's land	0.3	3.0	3.3				3.3
iii. Forest land	4.6	18.9	23.5				23.5
iv. Roadsides etc.	1.3	24.4	25.7	0.4	1.6	2.0	27.7
Total Collected	11.4	55.4	66.8	0.4	1.6	2.0	68.8
Purchased	8.7	3.3	12.0	11.1	2.6	13.7	25.7
Total	20.1	58.7	78.8	11.5	4.2	15.7	94.5

Leach, 1987

TABLE 2.7: **Sources of supply of woodfuels in 1991 in India**

Source	Details	Total Fuelwood Contribution in Million Tonnes/Year
Forests	Felling of trees	19.0
	Lopping of twigs and branches	42.0
	Logging wastes	9.5
Shrubs on degraded lands and roadsides		46.0
Tree planting on 17 million hectares during 1975-90 through social and farm forestry programmes	Tops, twigs, small branches and poles	40.0
Homestead gardens	Twigs and branches	16.0
Total		172.5

Ravindranath and Hall, 1995

These considerations make production of fuelwood by farmers for fuelwood market on non-viable proposition. One school of thought suggests growing such trees on public lands which produce lot of twigs and branches that can be gathered, and not through commercial production on farm lands.

Due to the popularity of growing trees on farm lands, the availability of fuelwood has considerably increased in certain regions of India. The shares of total collected fuelwood, represented by different sources of collection have also undergone a change as indicated in the Table 2.8.

TABLE 2.8: **Major source of collection of firewood (percentage of households)**

Items	1978-79	1992-93
Own farm	35.14	48.50
Roadside bushes and trees	23.90	29.80
Forests	35.42	17.00
Others	5.54	4.70
Total	100.00	100.00

Natarajan, 1996

On the other hand, nearly half of the households collect wood from their own farms in 1992-93. The share originating from forests shrank to half of the 1978-79 figure, mainly because of the legislation enacted by the State governings, banning felling of trees in forests in many parts of India.

Estimated Additional Supply Potential

In addition to the existing production of 172 MT of fuelwood, there is immense potential for increasing production of fuelwood by afforestation on degraded lands. N.C.Saxena has identified some of these lands as follows:

(i) Cultivated Lands

The area available for agro forestry and growing trees on cultivated lands is difficult to estimate. All cultivated land can grow trees, but tree growing will depend on the opportunity cost. Degraded lands can give lower productivity. Bhumbla and Khare 1984, estimated under sown land, subject to wind and water erosion, to be 38 mha. One third of it or 30 mha can certainly be utilised for additional agro forestry.

(ii) Strip Lands

These include farm bunds and boundaries, roads, railway lines, canals and drains. The Fuelwood Committee (Planning Commission, 1982) estimated potentials of these at 1.2 mha, 1,224,000 km of roads, 60,000 km of railway lines, 1,50,000 km of canals and 20,000 km of drains.

(iii) Degraded Forest Lands

The Forestry Survey of India (FSI, 1993) has estimated that almost half of the forest area has a crown density of less than 40 percent. The National Wastelands Development Board estimate that 36 mha out of the total of 57 mha of forest land are degraded (NWDB,1988: 28) and are, therefore, capable of growing additional trees which can provide fuelwood.

(iv) Uncultivated Degraded Land

Land statistics are well developed for cultivate and forest lands, but not for non-forest non-cultivated lands (Romm, 1981). Information needed for assessing their potential, like ownership, extent and type of degradation, and present and possible future uses, has never been collected on a systematic basis for the entire country.

TABLE 2.9: **Ownership of degraded land available for tree growing (mha)**

Category of cultivable Land	Total area	Available for trees (for new plantation)	Private	Forest Dept.	Revenue/ other depts.
Cultivated	142	13	13	—	—
Forest	67	36	—	36	—
Uncultivated/ non-forest	55	33	21	—	12
Strips	Included in the above	2	1	—	1
Total	264	84	35	36	13

Chambers *et al.*, 1989.

1.3 Main uses of fuelwood energy

Fuelwood is mainly used for domestic cooking and heating purposes. About 30% of the urban population and the total rural population uses fuelwood for their domestic needs. In addition some of the industrial or commercial activities also utilised fuelwood as an energy source. Details of biofuel supply metrics for the rural and urban areas is given in Table 2.10.

2.1 Monetisation of wood energy markets in rural and urban areas

The fuelwood prices showed upward trend during 1973-85. The real price of fuelwood increases by 34% in major cities of India during 1979-82 and another study of 41 towns showed 0.5% increase during 1977 to 1984. Same study compared movement of fuelwood prices with those of other commodities and observed that the price increase in fuelwood was 85% higher than that of wheat and rice. The annual growth rate average for 41 urban centres was 14.37% (Bowonder *et al.*, 1986)

TABLE 2.10: **Biofuel supply matrix: India 1978-79 (million tons a year)**

	Firewood logs	Firewood twigs	Crop resid.	Dung	Total
RURAL					
TOTAL	20.11	58.75	29.53	66.76	175.15
Purchased	8.65	3.29	.38	1.49	13.81
Collected	11.46	55.46	26.15	65.27	161.34
Collected from:					
Own land	5.24	19.06	17.56	51.63	93.49
Neighbour's land	.29	3.04	11.59	—	14.92
Forest land	4.65	18.93	—	—	23.58
Roadsides, etc	1.28	24.43	—	13.64	29.35
URBAN					
TOTAL	11.50	4.20	1.09	4.95	21.74
Purchased	11.06	2.55	.18	2.33	16.12
Collected	.44	1.65	.91	2.62	5.62
Collected from:					
Own land	.20	.57	.64	1.80	3.21
Neighbour's land	.01	.09	.27	—	.37
Forest land	.18	.56	—	—	.74
Roadsides, etc	.05	.43	—	.82	1.30
Total	31.6	62.9	30.6	71.7	196.8

Source: Natarajan (1985).

The price of fuelwood in major urban centres in India during 1984 and the size of major urban areas, based on population, gave an interesting reading. In all large urban centres, i.e., with a population of more than one million, the price of fuelwood was more than Rs. 500 per tonne. In all large metropolitan cities with a population of more than 4 million, the price of fuelwood was more than Rs.700 per tonne. Average price of fuelwood for 41 urban centres in 1960 was only Rs. 59.90 per tonne, whereas in 1984 it was Rs. 588 per tonne. In 1984, in Bombay, the price of fuelwood was Rs. 1,053 per tonne and in Delhi it was Rs. 950 per tonne. The annual growth rate of fuelwood prices was higher during 1977-84 than during the period 1960-1977. In other words, the rise in fuelwood prices was significantly higher during the period of 1977-84 when the gap theory was in currency. The sharp increase in prices was attributed to increased urbanisation, shortage of fuelwood and increased monetisation of fuelwood markets. Another important reason given was the reduction in forest area around the major urban centres.

The steep rise in the prices of fuelwood affected the economics of cooking and the reason for this is the efficiency of fuelwood stoves was about 10% , whereas the efficiency of kerosene stoves was about 40%. From the efficiency point of view, fuelwood is not the cheapest source of energy and the comparative advantage of fuelwood over kerosene can be reversed. Bowonder also observed that in 1984 (real terms) fuelwood was 2-3 times more expensive than kerosene for every unit of energy delivered to the consumer, whereas in 1960 for every unit of energy delivered, fuelwood and kerosene cost about the same. In short, in 1984, fuelwood was definitely more expensive. In Bombay, every user of fuelwood had to pay 4-5 times more for energy than if he used kerosene, and in Delhi 3.97 times, but in some centres such as Balaghat, Darjeeling and other places, fuelwood was still cheaper. The fact that fuelwood was more expensive to use than kerosene had immense implications for natural resource management.

The cost of fuelwood in urban areas has been a significant factor as there has been a sharp rise in number of people living in squattered settlements. It is estimated that 30% of the total population in the urban centres lives in such squattered land and a majority of them use fuelwood.

In the rural areas, the fuelwood has mostly been collected from the forest land or other lands and normally the price is not of significant importance. Certain projections in the cost of fuelwood in rural areas have been made by different organisations as follows:

Source	Area	Price (Rs./Kg)
1. Rural Energy Database TERI, 92	All India	0.5
2. Rural Energy Database TERI, 92 (South India)	Zaheerabad, Dt: Medak, A.P.	0.025 - 0.030
3. Rural Energy Database TERI, 92 (North India)	Barnala Taluka Dt. Sangrur, Punjab	0.60
4. Chambers *et al.*, 1989	Hyderabad, AP	0.61
5. Chambers *et al.*, 1989	West U.P.	0.30
6. Application of a Rural Energy Model in Engineering Planning, TERI, 1980	Haryana	0.25 - 0.70
7. Application of a Rural Energy Model in Engineering Planning, TERI, 1980	Rajasthan	0.15-0.30

Only 10% of fuelwood, purchased in rural areas. 90% collected by different sources at zero cost.

Despite the fact that degradation of forest land has continued and unabated and social forestry on community land has not been a visible programme, fuelwood prices in India after 1985 have not gone up for two reasons. As a matter of fact, fuelwood prices in Haryana in 1994-95 have been reported to have not only stabilised, but decreased after commercial plantation of fuelwood in different areas. (Narayan Singh, 1995). The two important reasons for prices not going up are:

i. Farmers having surplus land produced a great deal of eucalyptus wood. It was also sold as fuelwood being surplus to the needs for pols and pulp wood.
ii. The greatest potential for supply of fuelwood at a little opportunity cost is 2/3rds. These positive developments though unconnected with government policies, still leave out a large portion of rural population for home fuelwood is scarce.

3. EVALUATION OF STATUS QUO

3.1 Critical shortages/surpluses

At present no serious shortage of woodfuel supply is experienced by the rural or the urban populations. However, the availability or the accessibility is sometimes constrained due to its monetisation. In the areas where fuelwood is now commercially grown, its trading has made it mandatory that a certain price be paid for the fuelwood. Those who can afford are able to fulfil their demands appropriately but in certain cases, where people are not able to pay they will find it slightly difficult to access. The low socio-economic status of most traditional fuel users puts them in the position of free collectors of woodfuel, rather than consumers of traded fuel. Recent observation by NCAER indicates that with the improvement in technology both in terms of cook stoves as well as for the gasifiers, consumption of hard wood is on the increase. A good sign is that the efficiency improvements in the enduse devices has contained or arrested the overall requirement of the fuelwood.

There has been a gradual shift of rural population, from rural to urban areas and with them the change in the fuel requirement in the urban centres. Such a change comes in seasonally when the agricultural labour shifts to nearby urban towns in search of job and secondly in times of hardship or droughts the rural populations have a sudden high influx into the urban areas. These sudden influxes do bring around a shortage condition, which is not of a permanent nature. The problem that is acutely faced in such situations is that the people who shift mainly depend on the twigs, branches and tree felling, which all come free of cost. But these are not available in the urban centres, where one has to use the traded biofuel.

With the privatisation and commercialisation of tree farming, local availability of fuelwood is constantly on the increase. This fuelwood is penetrating into the local markets which used to receive the supply earlier either through the vendors or government established corporations. The government run agencies procured the fuelwood they supply from government forest department. Many vendors also acted as collectors, transporters and sellers of the informally traded fuelwood, and they mainly service the household sector. For most of the domestic fuelwood consumers, who are primarily the rural people and the poor in larger urban centres, fuelwood is still a free source.

It is thus observed that a gradual shift in the consumption pattern has made the wood trade a profitable business and larger proportion of fuelwood is now coming through the monetised channel. There are no perennial shortages.

3.2 Factors contributing to unsustainable wood energy management practices

There are few identified sectors, related to social environmental and financial aspects that become a problem sometimes in terms of sustained fuelwood supply. Fuelwood will remain an important sub-sector in the overall energy scenario. The consumption of fuelwood and other biomass fuels will increase in foreseeable future. Non-forest land will continue to be the main sources of fuelwood and as such market sources will slowly become active and monetise the fuelwood trade. Fuelwood consumption in general will not be the main cause of deforestation. Whereas the overall availability is foreseen there are certain policy areas that has its own impact. Some of the issues are as follows:

(i) Management of trees and forest resources by villagers, including the value addition on site through processing and marketing support is not available at the required level. This calls for policy support in the forestry sector.

(ii) Poverty elevation is a major concern and tree growing has been traditionally a source of income and subsistence for the rural poor, which is now getting shifted out from the forest lands and the rural areas to semi-urban areas.

(iii) Environmental concerns, globally have focused too much of attention on degradation of forest sources due to unsustainable consumption pattern of fuelwood.

(iv) Due to non-monetory reflections of the fuelwood supply. Its reflection in the national economic growth has not been appropriately reflected.

Supply of fuelwood from farm Land

Non-forest lands are a significant source of fuelwood. Production of fuelwood from farm lands can be considered under two heads:

i. Traditional agro-forestry; and home guarded models; and
ii. New activity of farm forestry.

Under the traditional agro-forestry pattern, trees are protected, planted and managed on farm lands in India in a variety of situations.

TABLE 2.11: **National aggregates of biofuel use in rural domestic sector**

(in million t/y)

	Firewood	Dungcake	Agricultural residues
Consumption			
NCAER	93.3	83.2	36.7
REDB (low)	181.3	40.1	31.6
REDB (average)	252.1	106.0	99.2
REDB (high)	309.4	114.5	165.5
IREP	169.0	54.2	62.8
Requirements			
ESI (510 kcal/pc/d or 2.13 MJ/pc/d)	151.3	41.5	47.8
EDSG (520 kcal/pc/d or 2.18 MJ/pc/d)	201.7	55.3	63.7
ABE (620 kcal/pc/d or 2.60 MJ/pc/d)	240.5	65.9	76.0

From Rural Energy Database, Page 4.18

TABLE 2.12: **Average fuel consumption based on national surveys: GJ/year**

	Per Person		Per Household	
	Rural	Urban	Rural	Urban
TOTAL ENERGY				
India 1979	4.80	3.76	28.22	21.63
TOTAL MODERN FUELS				
India 1979	0.25	1.35	1.47	7.80
TOTAL BIOFUELS				
India 1979	4.55	2.41	26.75	13.83
FIREWOOD & TWIGS				
India 1979	2.34	1.77	13.78	10.17

From Household Energy in South Asia Book, Page 12.

In Kerala, a region of high rainfall and good soils, farmers plant trees on home steads and on farms to maximise overall returns from land. In arid Western Rajasthan, farmers protect prospis cineraria and zizyphusspp trees to increase soil productivity and land sustainability. These trees recycle nutrients and provide both fodder for cattle and mulch and shade for crops, and these component farm production. In the hills, trees are maintained on farm boundaries for subsistence products like fodder and fuelwood. Casuarina plantation for urban fuelwood has been a part of the rural scope in southern coastal India for more than a century now.

Fuelwood from Non-forest Public Lands

Village, road sides, tank foreshores and other such open excess lands are the traditional sites for collecting fuelwood among rural people. Often, these degraded lands offer only woody shrubs. These are not favoured species, but the absence of commercial interest in these species help the poor in their access to such shrubs. In many semi-arid regions, the natural spread of prospis shrubs provides excellent fuelwood both for consumption and sale at almost zero opportunity cost to the poor. According to a field study of Prof.Ravindra Nath, 1995, in five villages of the semi-arid district of Anantapur (Andhra Pradesh) 86% of households met more than 75% of their cooking needs from prospis shrubs alone. Here, gathering of fuelwood from degraded public lands has become a cottage industry, in as much as of it goes to the nearby metropolitan town Bangalore. As much as 10% of the local population gets employment from this activity.

FUELWOOD PROBLEMS IN INDIA

Fuelwood problems in India or the gap theory is totally misleading. Land production capabilities and access to almost vary from region to region and this is an impact on gathering energy for cooking is a problem for people in the rural areas. One way to classify regions with respect to fuelwood availability has been recommended as follows:

i. Regions with proximity to forests;
ii. Fertile and irrigated and cultivated land;
iii. Areas with excess shrubs;
iv. Areas where farm forestry has been successful; and
v. Areas where dung must be returned to fields to maintain productivity.

It has been observed that fuelwood is an acute problem more in the last type of region, where it may cover half of India's geographical area. In forest regions the issue is not of physical scarcity, but of lack of income, which drives poor and land less may even collect shrubs, when it is not an issue for the surplus farmers.

Even if the poor have land, their immediate pre-occupation is the need for quick solutions to desperate food and income deficiency, and they cannot be expected to use their lands for the production of fuelwood.

DEMAND FOR BIOFUELS

In a survey conducted by the National Council on Applied Economic Research, NCAER, 1978-79, it was estimated that bio-fuel accounted for over 90% of the heating energy used by rural households. The major bio-fuel used were fire wood, crop wastes and dung cakes. Firewood along accounted for more than 55% of the total bio-fuel consumption. In another survey conducted ten years later in 1980, the energy demand has grown due to both the income and population increase and all the available information show that the rural household continue to depend heavily on bio-fuel to meet their energy requirements.

During these years there had been an impressive increase

in the availability of LPG and Kerosene, the two petroleum fuels essentially used in the household sector. This apparently had not made any dent in the pattern of fuel consumption in the rural areas. The Council had conducted two urban surveys in the last ten years to estimate the potential of LPG in urban India. The results of these studies show that the increased availability of Kerosene and LPG was mostly absorbed in the urban areas. (the two studies referred to NCAER, demand for LPG in urban areas, 1985 and demand for LPG in Northern region, 1995).

In 1993, the NCAER completed the study evaluating a national programme of improved chullah. The findings of this study was based on data collected from samples of 16,000 households, spread over 1600 villages selected from all the districts in the country. As a part of this study the data was collected from the sample households on the energy use and the pattern of fuel consumption. Based on these data the estimate of rural energy consumption and the pattern of fuel used was made and these were presented in a paper.

According to the above mentioned study, the estimated consumption of energy for heating by the rural households in 1992-93 was 153.2 million tonnes coal replacement, this is about 72% more than what was used in 1978-79. The pattern of fuel consumption for these two years are given in Table 2.13.

It can be seen that the share of bio-fuel in the two years has remained unchanged at about 95%. Fuelwood consumption has gone up both in quantity and share. The interesting feature is the increase in the use of logs in 1992-93 compared to the earlier years. The survey also showed that more people collected logs in 1992-93 compared to 1978-79.

NCAER also used this competitive date to forecast the energy consumption during 1999-2000. The estimated average household energy demand for another few years the income shift was considered the major component. In order to gauge the affect of a rise in real (household income) on average

TABLE 2.13: **Energy Consumption for Heating in Rural Households**

Fuel	Unit	1978-79 Qty.	CR	Percent	1992-93 Qty	CR	Percent
Coal/Soft coke	'000 t	1,143	1,713	1.92	429	584	0.38
Kerosene	'000 kl	414	2,279	2.55	1,103	6,831	4.44
Dung cake	'000 t	66,755	20,087	22.51	86,732	26,151	17.00
Firewood logs	'000 t	20,109	16,905	18.95	57,956	49,989	32.49
Firewood twigs	'000 t	58,742	31,774	35.62	73,418	44,785	29.11
Crop	'000 t	29,529	15,531	17.41	24,955	20,530	13.35
Others			913	1.03		4,971	3.23
Total			89202	100.00		153,841	100.00

energy consumption, it was necessary to know the income elasticity of demand for energy. Between 1978-79 and 1992-93, the average household heating energy demand grew at 1.9% per annum. This growth may be attributed mostly to the increase in the annual average household income during 1978-79 and 1991-92. For making projections, it was presumed that period between 1992-93 and 1999-2000 also the average household energy demand would grow at 1.9% per year. This essentially means that the income effect has taken to be the same in the two periods in consideration.

It is likely that the income would grow faster in the coming years compared to the past. What is relevant is how the income distribution would get affected through this growth. If bulk of this increase accrues to the affluent where the elasticity for heating energy is likely to be low, the average household energy consumption may not grow faster than what was observed in the past.

POPULATION PROJECTION

The rural population of India was estimated at 628 million in the 1991 Census. The growth rate between 1981-91 was estimated at 1.8% per year. The present indications are that the average annual growth rate for rural population between 1999-2000 will be about 1.7%. The projected rural population in 2000 will be 744 million. The expected average household size in the rural areas is expected to be 5.7 and the estimated number of rural households will be 130.5 million.

With an average household heating energy demand of 1.50 kg of coal replacement demand for heating energy in rural household in 1999-2000 is estimated at 202.4 MTCR.

TABLE 2.14: **Demand for heating energy and likely pattern of fuel use in rural households 1999-2000**

Fuel	Unit	Qty.	Cr '000 t	Percent share
Coal/Soft coke	'000 t	400	600	0.30
Kerosene	Th. K.l.	1,700	10,540	5.21
Bio-gas	Mill.cu.m.	1,020	3,570	1.76
Dung cake	'000 t	93,000	27,900	13.79
Crop Wastes	'000 t	38,000	23,180	11.46
Firewood	'000 t	182,133	136,485	67.47
Total			202,275	100.00

Table 2.14 reflected the demand for heating energy and likely pattern of fuel use in rural household in the year 2000. The demand for dung cakes is estimated to be around 93 million tonnes, burnable crop wastes demand is estimated at 38 million tonnes. Firewood is projected at 182 million tonnes.

The methodology used for arriving at this figure is simple. The NCAER study observed that the share of logs had increased in the total firewood from 25.6% in 1978-79 to 44.3% in 1992-93. This trend may continue further and it is assumed that 1999-2000, logs and twigs will have equal share in the total firewood demand. In this assumption the demand for fuelwood for 1999-2000 could be around 182 million tonnes.

Mix of wood fuels by supply source (MT/%)

Source	Qty.(MT)	Percent
Ownland *	34.32	20.31
Forest	75.98	44.96
Purchased	13.44	7.95
Other Area +	45.26	26.78
Total	169.00	100.00

*Agriculture land, house yard etc.
+Roadside, Panchayat land etc.

Source: "Analysis of the IREP Project Document" by centre for Rural Technology & Development, IIT, Delhi for Planning Commission, GOI, 1991

Data based on 106 blocks survey reports, conducted under the IREP blocks upto 1990.

Rural Wood Energy Consumption (1991)

	Per Capita (Kg.) day	Total (MT/year)
NCAER *	0.41	93.3
REBD *	1.10	252.10
IREP*	0.74	169.00
ARID+	0.53	151.48

**Source*: Rural Energy Database, Tata Energy Research Institute, New Delhi, 1992. (Table 4.5, page 4.18)

\+ Based on studies conducted by Association for Research & Integrated Development, New Delhi under the programme "Countrywide Biomass Assessment Scheme Under National Programme for Demonstration of Taluka level Biomass Based Power Projects". The studies were conducted in 6 States covering 13 sub-divisions (Talukas) 40 blocks and more than 700 villages (1996-97).

Total Rural Population 628.7 m—Census of India, 1991
Fuelwood consumption in India/year (1991)

	Rural	Urban
	(i) *Domestic*	
Per Capita (MT)	252.00	30.00
	(ii) *Industry*	
Per Capita (MT)	6.00	10.00
	(iii) *Total*	
Per Capita (MT)	258.00	40.00

Source: Biomass, Energy and Environment, Ravindranath, N.H. & Hall, D.O. Oxford University Press, 1995.

3

RENEWABLE ENERGY UTILIZATION IN INDIA

INTRODUCTION

Renewables, both in traditional form as well as in the modern form, are finding increasing role in the socio-economic sectors. Whereas the contribution in percentage terms from renewable energy sector appears to be decreasing because of the rapid growth of commercial sources of energy, their contribution in absolute terms is on the increase. Efforts made in India during the last two decades have given a well laid out policy and infrastructure facilities for developing the renewable energy sector.

Recent studies conducted by the Indian Association for the Advancement of Science and other similar institutions have indicated wider need for promotion of renewable energy through the route of modern conversion technologies. Awareness creation about the utility of such devices is at a very high level. Setting up a Ministry of Non-Conventional Energy Sources has helped to formulate a National Plan which gradually built up infrastructure in different federal states of India. Possibly, India is the only country which has seen different phases of technology development and dissemination.

Paper presented at the UNDP/ESCAP PACE-E REED-2 (Sri Lanka) National Orientation Workshop on "Integrated Rural Energy Planning and Environmental Assessment", organised by the Government of Sri Lanka, at Colombo, Sri Lanka during 4-6 December, 1997.

The initial thrust had resulted in renewable energy systems being disseminated at a very fast pace, mainly through the government support and subsidies. The desire to reach out to the users was appropriately responded, but was constrained by the amount of subsidy offered and the total quantum of subsidy available from government's funding. Certain observations like mushroom growth of fly by night operators had a dampening effect on the growth pattern after partial or total withdrawal of subsidy on different systems. Withdrawal of subsidy also resulted in a sudden break in the growth of this sector. However, government's consistent policy interventions, duly supported by funding from international agencies, resulted in creation of an environment where private entrepreneurs tried to establish a niche market for the renewable energy systems. As the World Bank initially identified Wind Power and the Photovoltaic sectors through private entrepreneurs, there has been a sudden growth in wind machine installation mainly through the private entrepreneurs. Long term fund availability and the technical back-up has resulted in stabilising this sector of energy supply.

Growth of renewable energy systems is to be seen in this context. Availability of infrastructure, and government's policy to support level playing field to renewables will certainly help in faster growth of the sector. Following description will give an overview of energy policy and growth of renewables vis-a-vis other sources of energy and the future prospects.

ENERGY POLICY

A number of high level committees have been set up in the past by the Government of India to examine in detail the various issues and suggest measures for ensuring development of the energy sector for sustaining the economic growth and meeting the basic needs of the people. The recommendations made by these committees are kept in view while formulating the Five Year Plans. The 'strategy in energy sector' adopted for all Five Year Plans incorporates the changing needs of the economy and the society. It was felt desirable during the formulation of the Eighth Five Year Plan (1992-97) to adopt a long-term strategy which is consistent with the goal of sustainable development. This energy policy is being studied and a new policy statement is likely in next six months. That will respond to the present situation.

The short-term and the long-term priorities as outlined in the Eighth Plan document are given as follows :

Short Term

— maximise returns from the assets already created in the energy sector.
— initiate measures for reducing technical losses in production, transportation and end-use of all forms of energy.
— initiate action to reduce the energy intensity of the different consuming sectors of the economy and promote conservation and demand management through appropriate organisational and fiscal measures.
— initiate steps for meeting fully the basic energy needs of the rural and urban households, so as to reduce the existing inequities in energy uses.
— maximise satisfaction of demand for energy from indigenous resources.

Medium-Term

— initiate steps towards progressive substitution for petroleum products by coal, lignite, natural gas and electricity so as to restrict the quantum of oil imports to the 1991 level.
— initiate action for accelerated development of all

renewable energy sources, especially the available hydro-electric potential.

— promote programmes to achieve self-reliance in the energy sector.

— initiate appropriate organisational changes in the case of different energy sub-sectors consistent with the overall energy strategy.

Long-Term

— promote an energy supply system based largely on renewable sources of energy.

PRIMARY ENERGY ENDOWMENTS

Commercial Energy Reserves

Present proven recoverable reserves of different sources of commercial energy in India (in 1993-94) are as follows :

TABLE 3.1: **Recoverable Reserves of Commercial Energy in India**

i.	Coal	68,047 mt
ii.	Crude Oil	757 mt
iii.	Natural Gas	686 bcum

Ref : Planning Commission, Government of India, 1997.

Emerging technologies are likely to make higher amounts of recoverable reserves in the country.

Renewable Energy Potential

A resource survey has indicated large potential of the renewable energy sources. Table 3.2 gives an overall potential in the country in 1993-94 :

TABLE 3.2 : **Renewable Energy Potential in India**

Sources/Technology	Potential/Availability
Biogas	12 million Nos.
Biomass	17,000 MW
High Efficiency Wood Stove	120 million Nos.
Solar Energy	5 × 10 WHr/Year
Small Hydro	10,000 MW
Wind Energy	20,000 MW
Ocean Thermal	50,000 MW
Sea Wave Power	20,000 MW
Tidal Power	9,000 MW
Urban Municipal & Industrial Waste	1,000 MW

Ref : Planning Commission, Government of India, September, 1995.

The potential cannot be taken as exploitable target under existing technology know-how. However, projection of this potential may lead to new technology development in India.

PRIMARY ENERGY PRODUCTION AND CONSUMPTION PATTERN

Energy Production

Amongst the indigenously produced primary commercial fuels, the relative share of oil and gas has increased from 1.2 per cent in 1950-51 to 32 per cent in 1990-91; whereas the share of coal has declined from 98 per cent in 1950-51 to 63 per cent in 1990-91. The share of coal is reflected to be 67 per cent in indigenously produced primary commercial fuels in 1995-96, mainly due to lower crude oil production.

Trends in indigenous production of commercial energy, for the period 1960-61 to 1995-96, are given in Table 3.3.

TABLE 3.3: **Indigenous Commercial Energy Production**

Source	Unit	1960-61	1970-71	1980-81	1990-91	1993-94 (*)	1995-96 (*)
Coal & Lignite	mt	55.72	76.34	118.81	225.50	264.14	292.27
Crude Oil & NGL	mt	0.45	6.82	10.51	33.02	27.76	35.20
Natural Gas	bcum	—	1.33	2.17	16.55	16.86	22.63
Hydro-Electricity	MW (e)	1,900	6,400	11,800	18,800	20,400	21,000
Nuclear Power	MW (e)	—	400	900	1,500	2,250	2,250

Ref. : Planning Commission, Government of India — September, 1995.
Economic Survey of India, 1996-97.
* Provisional Figures.

Pattern of Consumption of Commercial and Non-Conventional Sources in the Primary Energy Sector

The pattern of primary commercial energy production and consumption in the country has undergone a change over the last four decades as is evident from the energy balance given in Table 3.4. The total primary energy consumption (including commercial and non-commercial) in the economy has increased from 3,752 peta joules (PJ) in 1953 to 13,524 PJ in 1993. In terms of oil equivalent units, the total primary energy consumption has increased from 89.8 million tonnes of oil equivalent (mtoe) to 323.01 mtoe during the same period. The share of primary commercial energy consumption has gone up from 28.4% to 63.2% during the same period. The primary commercial energy consumption has grown at a rate of 5.34% per annum as compared to a growth rate of 3.26% in the total primary energy. As far as final energy consumption is concerned, it has increased from 3,537 PJ or 84.48 mtoe in 1953 to 10,569 PJ or 252.44 mtoe in 1993 at an annual compound rate of growth of 2.77%. The rate of growth of final commercial energy consumption has been 4.81% during this period. In absolute terms, the final commercial energy consumption in the economy has increased from 852 PJ (20.35 mtoe) in 1953 to 5,595 PJ (133.63 mtoe) in 1993. The household sector (mainly rural household sector) accounts for the entire final consumption of non-commercial energy.

The change in the relative share of commercial and non-commercial sources in the primary energy supplies and the share of different sources in the final commercial energy consumption are shown in Table 3.4.

TABLE 3.4: **Change in Pattern of Energy Consumption (%)**

Year	Primary Energy		Final Commercial Energy				
	Commer-cial	Non-Commer-cial	Coal	Pet. Prod.	Nat. Gas	Elec-tricity	Total
1953	28.4	71.6	80.1	16.7	0.0	3.2	100.0
1960	34.7	65.3	75.3	19.9	0.0	4.8	100.0
1970	40.6	59.4	56.1	34.1	0.6	9.2	100.0
1980	47.5	52.5	47.9	40.3	1.1	10.7	100.0
1990	59.9	40.1	35.9	43.6	5.5	15.0	100.0
1993	63.2	36.8	32.0	44.5	6.4	17.1	100.0

Ref: J. N. Maggo — R&D Needs for Energy Sector — September, 1996

As may be seen from Table 3.4, the share of coal is declining in the final commercial energy consumption, and that of other fuels is increasing, particularly of oil products.

Table 3.5 gives the source-wise consumption of final energy.

Share of Different Sectors in consumption of commercial energy is given in Table 3.6.

Non-Commercial Sources of Energy (Biomass only)

Traditionally the biomass based renewable energy sources

TABLE 3.5: **Source-wise Consumption of Final Energy (Peta Joules)**

Year	Non-commercial Energy				Commercial Energy				
	Fuel-wood	Dung Cake	Agro-Waste	Total	Coal & Lignite	Pet. Prod.	Nat. Gas	Elec-tricity	Total
1953	1700	466	519	2685	683	142	-	27	852
1960	1968	543	603	3114	952	252	-	61	1265
1970	2345	653	633	3631	1070	649	12	175	1906
1980	2864	794	774	4432	1444	1212	33	323	3011
1990	3210	890	874	4974	1827	2218	283	762	5090
1993	3210	890	874	4974	1792	2490	357	956	5595

Ref : J.N.Maggo — R&D Needs for Energy Sector—September, 1996.

TABLE 3.6: **Share of Different Sectors in Consumption of Commercial Energy (%)**

Sector	1953	1960	1970	1980	1990	1993
Household	9.6	9.7	12.2	9.7	11.8	12.0
Commercial	0.9	0.3	1.2	0.9	1.1	1.2
Industry	39.9	41.5	49.6	49.6	44.7	42.8
Agriculture	1.4	1.6	2.7	6.3	8.2	9.2
Transport	43.7	40.2	28.2	23.5	19.8	20.8
Others	1.5	1.7	0.4	0.4	0.9	0.9
Feedstock	0.5	0.3	4.2	5.6	10.4	10.4
Non-Energy	2.9	4.7	1.5	4.0	3.1	2.7

Ref : J.N.Maggo — R&D Needs for Energy Sector—September, 1996.

contribute only to rural domestic energy supply. Other renewable energy sources contributed marginally in the overall energy scene. Therefore, only the biomass resources has been considered. Following Table 3.7 gives details of contribution of different biomass energy sources.

TABLE 3.7: **Biomass Use as Energy Fuel in India (1991 - 92)**

(mt/ year)

Sl. No.	Sources	Availability	Estimated Consumption	Deficit
I	Fuelwood	28.	166.85	138.40
II	Urban/Indl.Waste	246.0	Negotiable	_
III	Agro Residue	240.0	150.0	_
IV	Cattle Dung	960.0	80.0 (Dry Wt)	_

Ref : Planning Commission, Government of India—September, 1995.

RENEWABLE ENERGY

Background

The renewable energy programme was initiated in India on a very moderate scale in the wake of the energy crisis of 1973 when abundant endowments of renewable energy sources—solar, wind, small hydro, bio-mass, ocean thermal—gradually received attention of the energy experts and the policy makers. As a follow-up to the 1981 UN Conference on New and Renewable Sources of Energy, which underscored the need for drawing up plans for achieving a harmonious transition from an economy based on hydrocarbons to one based on renewable energy resources, the Government of India decided to set up a high-powered Commission for Additional Sources of Energy (CASE). A separate department was also created in 1982, under the Ministry of Energy, at par with Departments of Coal and Power, to provide focussed attention to this sector. After a decade, in 1992, the government decided to give a further boost to increase the deployment of near and already commercial renewable energy technologies by upgrading the erstwhile Department to a full-fledged ministry.

Under the New Strategy and Action Plan of 1993, the Renewable Energy Development Programme has been given market orientation, with major growth expected through the private sector and the non-government organisations.

Action Plan

A two-pronged Action Plan is as follows :

i. High priority has been given to generation of grid quality power from wind energy, small hydro power, bio-energy and solar energy.

ii. Rural energisation programme is promoted through stepping up of electrification of villages through photovoltaic and bio-mass gasifier power systems, supply of solar lanterns to unelectrified households, use of solar water heating systems and rural energy programmes viz., National Project on Bio-gas Development. National Programme on Improved Cookstoves and Bio-gas Programme and production of energy from industrial and agricultural waste.

Strategy

A three-fold strategy has been pursued by the government for promotion of renewable sources of energy through private sector involvement. These include :

(a) providing budgetary resources by government for demonstration projects.

(b) extending institutional finance from the Indian Renewable Energy Development Agency (IREDA) and other financial institutions for commercially viable projects, with private sector participation; and external assistance from international and bilateral agencies.

(c) promoting private investment through fiscal incentives, tax holidays, depreciation allowance, facilities for wheeling and banking of power for the grid and the remunerative returns for the power provided to the grid.

NEW THRUST TO RENEWABLE ENERGY PROGRAMME

A. Policy and Legislation

i. A comprehensive policy and a separate legislation for renewable application is being prepared.

ii. Creation of appropriate and empowered institutional structures in the States has been undertaken to deal with renewable energy.

B. Integration of Renewable Energy

i. Institutional linkages for integration of renewable energy being strengthened with major consumers of energy, and where possibility exists of substituting/or inducting renewables in their energy budgets.

ii. Major programme of solar PV power sources for rural telecommunication systems being promoted.

iii. An interaction mechanism being established with Rural Electrification Corporation to take up rural electrification through renewables. This is relevant for 18,000 villages, that cannot be otherwise electrified.

iv. States being asked to integrate renewables as a supply side option in their power plants. Independent power producers are also being motivated.

C. Identification of High Focus Areas

i. Distinctive features of the current renewable energy programme is to identify a limited number of high focus areas for concentrated deployment of systems. These are far away places, hilly areas and tribal villages.

Completely solarizing the 12,000 houses on Sagar Island is an example.

D. Marketing Outlets

i. Effort is being made to expand retail marketing outlets for renewable energy products. Aditya (Solar) shops being set up in major cities and towns with financial assistance from the government.
ii. Major NGOs being promoted to set up such shops.
iii. Trained technical persons, manning these shops, will also provide post-sale maintenance support.
iv. Cooperatives of renewable energy entrepreneurs for the sale, installation and service of renewable systems in small towns and rural areas is being promoted.

E. R&D and Specified Institutions

i. A fair amount of competence in bio-mass gasification built up with an R&D Centre at Walchand Nagar Industries.
ii. Solar Energy Centre is the major facility for testing solar thermal and photovoltaic instruments.

F. International Partnerships and Exports

i. India has extensive policy, technical, institutional and operational experience in renewable energy.
ii. India can offer, consultancy services, organise training programmes, transfer technology, set up plants for manufacturing of a whole range of renewable energy devices, and support demonstration projects in other developing countries on TCDC basis.
iii. India has launched a major drive to export products,

systems, technologies and services.

iv. To promote exports of hardware, technology and services a Standing Advisory Committee on Export of Renewables has been set up by the Government of India

RECENT ADVANCEMENTS IN USE OF RENEWABLE ENERGY SOURCES

Grid Connected Power Generation from Renewables

Contribution of renewable sources of energy is continuously increasing in the national sector. With technology development for conversion into improved final energy forms. A total capacity of 732 MW applications may be on a large scale. A total capacity of 732 MW of wind generators has been established. by the end of 1995-96, of which 684 MW has been installed in the private sector. Also a capacity of 121 MW has been created under the small hydro power projects, and 234 MW is under installation. The range under minihydro power has been winded to include systems from 10 KW to 15 MW. Large joint sector projects are being launched for power generation through solar thermal route. Photovoltaic applications have established their need for small scale electricity needs and the remote areas.

Energy from Urban and Industrial Wastes

A National Programme has also been launched since June 1995 for proper treatment of urban and industrial wastes and resultant recovery of energy from these sources. With the launching of this programme, an era of modern technologies has been initiated. Other applications of bio-energy include promotion of fast growing species, gasifier systems for power generation, briquetting of agro residues, bio-gas digesters and improved chulhas.

Solar Thermal Applications

Exciting innovations have been introduced in the solar thermal sector. Earlier, efforts at technology development has resulted in commercial development of market for solar water heaters, solar cookers etc. Commercial banks are also launching schemes for funding of such devices.

Solar Thermal Power Applications

A Solar Energy Enterprise Zone is proposed for the desert region in western Rajasthan for rapid commercialisation of various solar technologies through facilities and incentives for manufacture of equipment and installation of solar power generating capacity. A 140 MW integrated solar combined cycle project based on parabolic collectors and gas/steam turbine is planned for Jodhpur. The Government of Rajasthan is also taking action to set up solar power projects of 300 MW capacity in the State based on solar photovoltaics and solar chimney technologies.

Solar Thermal Water Heating

Solar thermal water heating systems are now coming to be increasingly used, particularly in commercial and industrial establishments. Over 325,000 square metres of collector area have so far been installed ranging from domestic water heater of 100 litres per day capacity to industrial systems of 240,000 litres per day capacity of water.

Solar Photovoltaic Applications

Solar PV is finding applications in decentralised fashion, especially for rural electrification and water pumping for irrigation, especially in dispersed hamlets of villages. India being the second largest solar panel manufacturer of crystalline silicon solar cells has a large manufacturing capability and technical back up support. Industrial production has touched a level of 8 MW per year as compared to the global production of 85 MW per year in 1995. Twelve different companies are engaged in the production of solar cells and modules and another 70 companies are involved in designing and supplying solar PV systems. Over 300,000 solar PV systems aggregating to about 22 MW have been installed.

Cooking Energy Equipment

In the area of cooking energy, 2.2 million families have biogas plants and 20 million improved wood stoves have been installed. Nearly 400,000 solar cookers are also in use.

The thrust of the governments approach has been to consolidate and strengthen technology development and promote commercialisation by providing a market orientation to various programmes. Therefore, financial support is being given for select demonstration projects, while creating attractive opportunities for private sector projects through a favourable policy environment and availabilityof institutional finance and external assistance.

FISCAL PROMOTIONAL MEASURES

Various measures for promotion of renewable energy sector have been tried by the government. Incentives to enable renewables to have a level playing field is considered essential. The Government has introduced a package of incentives for renewable energy projects which include: tax concessions such as 100% accelerated depreciation, soft loans, customs and excise duty reliefs and liberalised foreign investment procedures. Guidelines have also been developed and circulated to federal state governments on general policies and facilities for wheeling, banking, third party sale and purchase of power by utilities from such projects. Eleven states have announced their policies for purchase of power from renewables at the price of Rs. 2.25 per KWH. One state has also announced an annual escalation of 5% on that price. These policies are likely to motivate private developers of electricity from renewables.

The Indian Renewable Energy Development Agency Limited (IREDA), the financing arm of the government, is operating a World Bank line of credit amounting to US $ 195 million to promote private sector participation in renewable energy projects. Wind power generation and canal based small hydro projects, and end use solar photovoltaic systems are eligible for financing at low interest rates. These range from 2.5 to 5 per cent in case of photovoltaics systems to 18 to 19 per cent in the case of small hydro and wind power projects. A line of credit from the Asian Development Bank amounting to US $ 100 million has also been recently approved for operation by IREDA for assisting bagasse based cogeneration, solar thermal systems, bio-methanation and wind energy projects. Assistance from the Global Environment Facility, GEF, has been provided to the Ministry for hilly hydro and bio-methanation programme. USAID supports advanced biomass cogeneration programme.

FUTURE ENERGY DEMAND

Expert consultations at the Asia Energy Vision 2020, organised under the auspices of the World Energy Council, agreed on energy demand projection in India, upto 2020, as given in Table 3.8.

By the end of the Eighth Five Year Plan, i.e., 1992-97, renewables have a share of 1.5% of the total installed power generating capacity which is absolute term 1,200 MW out of a total of around 90,000 MW. There is keen private sector interest for the manufacture of renewable energy equipment, development of power projects and in the service sector. Also, it is noticeable that 7% of the total additional power generation capacity achieved during the Eighth Five Year Plan period has been contributed by the renewable energy sector. The trend in power generation from wind, micro hydal, and solar will add upto 10,000 MW by 2010 and 20,000 MW by 2020.

TABLE 3.8: **Future Demand for Commercial Energy in India**

Source		Unit	1991-92	1996-97	2009-10	2020-2021
1.	Electricity	TWH	231	336	725	1300 (1745)(2)
2.	Coal	mt	229	311	690(3)	1345
3.	Pet. Products	mt	57	81.2	165	335
4.	Nat. Gas	bcum	18.6	30.2	65	130

Ref: Planning Commission, Government of India — September 1995 & September 1996

1) Projections to 2020-21 made on the basis of business as usual. - U.N.Maggo - R&D Needs for Energy Sector - September, 1996.

2) Electricity generation capacity of 1745 TWH will be required to take care of T&D losses and Plant Load Factor. Corresponding installed generation capacity will be 378,000 MW at present day PLF and T&D loss level.

3) R.K. Sachdev - International Conference on Coal - June, 1996.

Projections have been made upto year 2020, as the present trend of primary energy dominance will be replaced by a shift towards renewable energy based systems—mainly biomass based. However, other modern conversion technologies will be suitably adopted for biomass. Power generation from; (i) fuelwood incineration, (ii) agro-residue briquette gasification, and (iii) cogeneration in sugar mills. Most conservative estimate indicates that power generation from bio fuels will have an established 20,000 MW capacity, that too in dispersed fashion. From 2020 onwards, additional capacities in power generation may be more from renewables.

Biomass may contribute to 12-14% of power generation.

The 25 years time span from now to 2020 provides sufficient breathing space to develop commercially viable technologies for bio fuels.

THRUST OF ASIA ENERGY VISION 2020

The final recommendations of Asia Energy Vision 2020 held during November 15-17, 1998 and which represent the South Asia Forum of the World Energy Council had focused attention on renewable energy supply forms. To quote specifically from the New Delhi Declaration 1996: "Rural poverty, and the need to create conditions for rural economic and social development, is the first need to be addressed. The necessary response is the installation of decentralised, locally appropriate and affordable renewable energy systems." This was defined as the prime goal in South Asia for attaining sustainable energy supply.

CONCLUSION

Renewable sources of energy have attained its niche market, and with the involvement of the private sector at all levels of industrial activity, the future for renewable energy systems in India is assured as reliable technical and manpower systems and infrastructure exist in the country. Financial support from banking institutions is a clear indication of bankability of renewable energy based projects.

4

POWER GENERATION FROM BIOMASS

INTRODUCTION

In India, about 74% population lives in rural areas. Rural electrification is the main factor for rapid development of rural areas. It helps in increasing the agricultural and village industrial production. Out of 26 States in India, 13 States have achieved 100% electrification. In the remaining 8 States, the level of village electrification is moving towards 100% albeit relatively slowly due to difficult and remote location. Rural Electrification Corporation (REC), a Public Sector Undertaking under the Ministry of Power has been implementing rural electrification programme through State Electricity Board (SEBs) and Rural Electric Cooperatives, since its creation in 1969. Though, the rural electrification programme has expanded fast in the last 30 years but there are over 86,000 villages, still to be connected to the grid. The four states of UP, Bihar, Orissa and West Bengal account for nearly 77% of the remaining un-electrified villages. The break-up of un-electrified villages in 13 States with less than 100% village electrification is given at Annexure-I. Most of these villages are located in hilly, tribal, desert, forest and ravine areas. A large proportion (50%) of the unelectrified villages are having population of less than 500; or about 80 to

Paper presented at the National Workshop on "Thrust Areas in R&D in Biomass for Sustainable Development", organised by the Bio Energy Society of India on July 11, 1998 at IIT, Delhi.

100 households (depending upon the size of family units). In most of the villages conventional systems are considered uneconomic.

In view of techno-economic non-feasibility of about 40,000 villages located in hilly and deep forest, desert, ravine areas there is need to find acceptable least cost option to electrify. Renewable energy sources such as solar, wind, biomass, small hydro etc have been considered as viable option for production of electricity in decentralised and sustainable manner.

POWER GENERATION FROM RENEWABLES

Mr. A. Parthasarthy, Secretary, MNES, in his address to the Asia Energy Vision 2020 on November 17, 1996 highlighted the renewable energy based power generation programme in India.

There is a significant potential in India for generation of power from non-conventional energy sources such as wind, small hydro, biomass and solar energy. In the Eighth Plan (1997-2002), special emphasis was given to the generation of grid quality power from renewables. The Government has given a market orientation to the programmes to make them commercially viable and sustainable. The limited budgetary allocations are being utilised for selected demonstration projects. Against the target of 600MW, a capacity of nearly 1100 MW has already been added and a total capacity addition of 1400 MW is expected to be achieved by the end of the Eighth Plan. Most of this capacity has come about through commercial projects.

The wind power potential of India is estimated to be at least 20,000 MW. Considerable progress has been made in harnessing this potential. India now has the third largest wind power installed capacity in the world, which has reached 820 MW. Wind power has been recognised as one of the "crowning achievements" of renewable energy development in

India. To support this, effort, MNES is also implementing the world's largest wind resource assessment programme.

The potential of small hydro power in the country has been assessed at about 10,000 MW. A total capacity of 133 MW has been constructed and projects of 246 MW capacity are under construction. The private sector has shown keen interest in small hydro and are bidding actively against the 600 MW capacity that has been offered in a few States. The Ministry is also implementing programmes for decentralised energy supply through community participation under the UNDP/GEF Hilly Hydro Project and by promoting portable micro hydel sets.

A major initiative in the area of solar power is the proposed Integrated Solar power is the proposed Integrated Solar Combined Cycle Power Project of 140 MW capacity near Jodhpur in Rajasthan. The World Bank/GEF and the German Government have agreed to provide assistance for this project. Work on the project is likely to commence next year after the completion of the bidding progress. The Government of Rajasthan has also awarded a Letter of Intent for a 200 MW Solar Chimney Project in the State. Under a joint initiative of the Ministry and the State Government, the establishment of a Solar Energy Enterprise Zone in western Rajasthan is being planned to promote large-scale solar power generation. The solar photovoltaic programme is being expanded to cover grid-connected voltage support systems on weak, rural grids and roof-top 'peak shaving' systems in urban centres. Two experimental power projects of 100 KW capacity each have been completed in Uttar Pradesh. The Rajasthan Government has awarded Letters of Intent for two solar photovoltaic power projects of 50 MW each. The UP Government is also planning to promote large commercial solar power projects in the State.

During the last few years, the Ministry of Non Conventional Energy Sources has created new opportunities for power generation from woody biomass and agricultural residues. The total potential is estimated at 17,000 MW. A new programme on bagasse-based cogeneration has been launched, which aims at optimising the generation of surplus power for the grid from the sugar mills. As much as 55 MW surplus power generation capacity has already been established and projects aggregating to 74 MW are under construction. The capacity of about 250 MW is under finalisation in the major sugar-producing States. New initiatives have also been taken in biomass combustion and taluka-based biomass power plants. Three private sector biomass power projects are at an advanced stage of construction and a capacity of about 40 MW will be operational by the end of this year (1997). A USAID/GEF project on Alternative Biomass Cogeneration is being implemented to enhance energy generation through the use of other biomass materials.

Through these various initiatives, grid quality power generation from renewables has attained maturity during the Eighth Plan. By the end of this Plan, around 7% of the total addition to the generation capacity achieved during the Plan is expected to come from renewables. Renewables are expected to have a share of 1.5% of the total installed power generation capacity by the end of this Plan—around 1400 MW out of a total of around 88,000 MW. There is a keen private sector interest for the manufacture of renewable energy equipment, development of power projects and in the service sector. The total annual turnover in this sector has reached about Rs,2000 crore.

Subsequently in an announcement on February 20, 1998, the Ministry of Non-Conventional Energy Sources informed the following:

India is well-endowed with substantial renewable energy resources estimated at around 50,000 MW. The exploitation of the potential depends upon the status of technology, availability of land and the end-use. The Ministry of Non-Conventional Energy Sources has chalked out a three pronged strategy for the promotion of new and renewable energy sources. These include:

(a) Providing budgetary resources for demonstration projects by State Governments, direct capital subsidy to State Electricity Boards and interest subsidy to industry including financial institutions to soften the loans for facilitating the commercialisation of renewable energy projects.

(b) Extension of institutional finance from the Indian Renewable Energy Development Agency (IREDA) and other financial organisations for commercially viable projects;

(c) Promoting commercial investments through fiscal incentives including tax holidays, accelerated depreciation allowance, facilities for the wheeling and banking of power in the grid and remunerative price for the power provided to the grid by commercial renewable energy projects.

A comparison of the earlier two statements indicates that biomass based power generation is picking up.

Note: MNES announcement of February 20,1998 has been added to the paper to indicate current status and approach.

BIOMASS POWER GENERATION PROGRAMME

The biomass power generation activities in India cover the following:

(a) optimum bagasse-based cogeneration for surplus power generation.

(b) direct biomass combustion-based power generation .

During the last three years of government supported

programme implementation, 18 biomass power projects with a total capacity of 69 MW have been installed in five States, which are feeding power to the grid and other 17 projects in seven States aggregating to 97 MW are under installation. Over 100 million units of electricity have been fed to the State grids so far from these projects. After commissioning of the above mentioned capacity, more than 800 million units will be fed to the grids every year, saving 500,000 tonnes of coal and reducing emission of 1 million tonnes of carbon dioxide into the atmosphere.

Bagasse-based Cogeneration for Rural Electrification

National Programme

The National Programme on Bagasse-based Cogeneration was launched in January, 1994 for the promotion of surplus power generation in the country's sugar mills, for which a potential of around 3500 MW was estimated. The programme has three main components, namely, (i) demonstration, (ii) interest subsidy, and (iii) financial assistance for preparation of detailed project reports. In addition, provisions exist for support to R&D, awareness creation, training, etc. During the year 1996-97 certain modifications were brought about in various provisions of the demonstration and the interest subsidy schemes with a view to making them still more attractive for the sugar mills to adopt cogeneration for surplus power generation.While the revised demonstration scheme provides for special dispensation to sugar mills in cooperative/ public sector with considerably enhanced financial support, the level of grants under the interest subsidy have also been enhanced significantly. The programme now covers cogeneration projects of almost all categories, (i) those employing advanced technology; (ii) those employing commercially established technology, as well as (iii) those which only carry out minor adjustments in their existing systems for small surplus power generation.

Different components to programme are defined below:

(i) *Demonstration Scheme* — This scheme provides for capital subsidy to 12 projects distributed among the major sugar producing States. Nine of these projects are to be taken up in cooperative/public sector mills, for which MNES support will be upto a maximum of Rs. 2.0 crore (20 million)/MW of surplus power generated. This amount will comprise of subsidies and soft loans. A set of eligibility criteria has been specified for these projects. The scheme is intended to promote model projects which can be replicated by other mills. For projects in the private sector or those taken up by IPPs, the MNES support remains unchanged at Rs. 70 lakh (7 million)/MW of surplus power.

(ii) *Interest Subsidy Scheme*—This scheme provides for capitalised grants of upto Rs. 35 lakh (3.5 millions)/MW to the leading financial institutions for projects of more than 4 MW of surplus power capacity to enable the institutions to reduce interest rates on their term loans for cogeneration projects. Reduced amounts of grants are also available for other smaller cogeneration projects using second hand boiler or turbine at half the level of grants.

(iii) *Financial Assistance for DPR Preparation*—A grant to an extent of 50% of the cost of preparation of detailed project report on the basis of certain benchmarks is provided to those projects which fulfil a set of eligibility criteria. The assistance is linked to surplus power proposed to be generated.

Ongoing Activities

The three initial projects in cogeneration have been launched in Madhya Pradesh, Andhra Pradesh and Tamil Nadu. In Madhya Pradesh, a 5 MW rice husk based project works an FBC technology. 12 MW project in Tamil Nadu, which will use cane trash and bagasse as fuel. A total of around 12,000 tonnes of these materials will be required. When commissioned, this will be one of the first few projects to use cane trash in any significant quantities. The project is expected to be commissioned by April 1997. These three projects are expected to serve as precursors to many more such projects for the utilisation of surplus biomass materials.

Under the Biomass Assessment Scheme, 96 studies have been undertaken through the State Nodal Agencies in the States of AP, UP, Gujarat, Assam, West Bengal, Punjab, Haryana, Madhya Pradesh, and Tamil Nadu. MNES is providing a grant of around Rs.1.0 lakh for each study. State-wise details of studies are given in Table 4.1.

TABLE 4.1: **State-wise List of Biomass Assessment Studies**

S.No.	State	Nodal Agency	No. of studies
1.	Andhra Pradesh	NEDCAP	12
2.	Madhya Pradesh	MPUVN	12
3.	Punjab	PEDA	10
4.	Maharashtra	MEDA	12
5.	Gujarat	GEDA	12
6.	Tamil Nadu	TEDA	10
7.	Assam	ASCS&T	03
8.	Haryana	HCS&T	04
9.	Uttar Pradesh	NEDA	10
10.	West Bengal	WBRDA	03
11.	Karnataka	KCS&T	05
12.	Arunachal Pradesh	APEDA	03
	Total		96

Village Electrification through Biomass Gasification

Biomass is a locally available and very convenient fuel for generation of electricity in rural areas. The most developed process is gasification. Biomass gasification is basically conversion of solid biomass (i.e. wood/wood waste, agricultural residues etc.) onto a combustible gas mixture normally called "producer gas" (or low Btu gas). The process is normally used for woody and powdery biomass and it involves partial combustion of such biomass. Partial combustion process occurs when air supply (or, more precisely, oxygen) is less than adequate for combustion of biomass to be completed. Given that biomass contains carbon, hydrogen and oxygen molecules, complete combustion would produce carbon dioxide and water vapour. Partial combustion produces carbon monoxide as well as hydrogen which are both combustible gases.

Producer gas thus generated after proper cooling and cleaning can either be used in burners, furnaces or can be fed to diesel generator in dual fuel mode for generation of mechanical or electrical power. Biomass gasifiers have been indigenously developed and commercialised for generation of mechanical, thermal and electrical energy in the capacities ranging from few kilowatts to several megawatts. Gasifiers are modular in design and are appropriate for generation of power in decentralized manner from locally available or grown biomass. Power generated through gasifiers can be used either for captive use or village electrification.

The initial efforts for village electrification through biomass gasifiers were made under MNES supported R & D projects at IIT, Delhi and IISc. Bangalore. IIT, Delhi, in collaboration with Himalayan Environmental Studies and Conservation Organization (HESCO), Kotdwar, UP had identified six unelectrified villages of Pauri Garhwal district and installed six biomass gasifiers of 3.5 KW each in 1992-93. Lantana Chamera, a wild problematic weed growing in that area has been used as feed material after charring. Electricity generated was being used for lighting of homes and streets. The harvesting of Lantana Chamera has also spread to land for agriculture use in these villages.

IISc. Bangalore, under MNES supported Gasifier Action Research Centre installed a 5 KW biomass gasifier in unelectrified village of Hoshahalli, of Tumkur district of Karnataka in 1989. The evaluation study carried out by IISc., Bangalore for Hoshahalli gasifier project has concluded that the payback period of the system is 4.82 years and 3.53 years for 3.5 KW and 20 KW system respectively. The unit cost of electricity has been estimated to be more than diesel generated power.

Since 1993-94 village electrification has been included in biomass gasifier demonstration programme, implemented by MNES in all States/U.T. Five villages in Bihar, Orissa, Karnataka and West Bengal have been taken up for electrification through gasifier. The largest project on installation of 500 KW biomass gasifier for electrification of Gosaba village in Sunderbans, in West Bengal has been sanctioned during 1995-96 and is operational.

Case Studies on Major Projects

Following Case Studies give details of these Projects:

The case studies of two projects taken up for electrification of villages of Hoshahalli and Gosaba through biomass gasifier are mentioned below:

Hoshahalli Project

Village Hoshahalli is located at a distance of about 110 kms. from Bangalore, in Karnataka State. The village population is about 250 and was unelectrified till 1968. The farmers' income is mainly from growing mulberry crops as the area is silk producing. In 1988, under MNES supported research and development project, Indian Institute of Sciences (IISc) decided to carry out plantation on 2 ha of community land, and installation of 5 KW gasifier in the village. The system was replaced by a 20 KW gasifier in 1994 to meet the increased demand of electricity of the villagers. The local people participated in the project actively and provided land for gasifier installation. The details of the project are given below:

— Gasifier	:	20 KW (IISc.design)
— D.G. Engine set	:	2×15 KVA, 3 phase
	:	1500 rpm (Kirloskar make)
— Wood Consumption	:	20 kg./hr
— Daily hours of operation	:	6
(a) Domestic	:	40 W fluorescent lamp and 15 W incandescent lamp in each of 43 houses
(b) Community	:	8 street lights of 40 W fluorescent lamp 3 hp pump
(c) Village Industry	:	7.5 hp flour mill
(d) Irrigation	:	2 pumps of 3 hp each for mulbery fields
(e) Management	:	The system has been managed by local people.
		Villagers pay Rs. 20 per month for lighting and drinking water supply.
— Cost of the project	:	Rs. 3.50 lakhs approx.
— Cost of electricity Generation	:	Rs. 3.00 per KWhr. approx.
— Other benefits	:	The living conditions of the villagers have improved a lot.

Gosaba Island Project

Gosaba is an island village in Sunderbans and is approachable by boat/barges from Sonakhali which is about 100 kilometers from Calcutta. The main village has population of about 16,000 and has no electricity; but, has substantial need.

About 100 ha of foreshore land is available for active energy plantation near the village. There is an existing forest on about 10,000 ha land by the forest department and social forestry division of Sunderban Development Board. The plantation techniques have standardized successful plantation consists mainly of a combination of bean, Kanakra, about 1000 ha of forest and social forestry control by the forest department. Government of West Bengal, 25% of the yield is committed for local people/panchayat and remaining 75% is sold to the outsiders at a price of Rs.300-400 per ton.

In view of the appropriateness of site for taking up dendro power project, a 500 KW (5×100 KW) gasifier for power generation (electrification) has been taken up at Gosaba village with financial support (75% of project cost) from Ministry of Non-Conventional Energy Sources, Government of India. West Bengal Renewable Energy Development Agency (WBREDA), Government of West Bengal are executing the project in cooperation with State Forest Department which has agreed to supply the wood for the gasifier. At a later stage dedicated energy plantation will be taken up by the local panchayat. Government of West Bengal is contributing 25% of the project cost. The details of gasifier power generation system are given in Table 4.2:

RESEARCH AND DEVELOPMENT

After taking up the implementation of the programmes for promotion of biomass-based power generation technologies, the Ministry has made conscious efforts to identify areas through which the potential for power generation in each of the technologies could be improved and costs could be brought down. Through interaction with the users of the technologies, such sugar mills and with academic institutions, an initial

TABLE 4.2: **Gasifier Power Generation System**

— Total installed capacity	:	500 KW (5×100KW)
— Gasifier capacity	:	5×100KW grassfiers of Indian make
— Diesel - Generator sets	:	5×125 KVA, 1500 rpm generator
	:	sets of Indian made using MWM turbo charged engine and alternators.
— Total cost of the plant including civil works, transmission and distribution	:	Rs. 90 lakhs
— Expected annual power generator	:	2.2 million KWhr
— Expected annual cost of wood @ Rs.400/ton	:	Rs. 10 lakhs
— Expected annual cost of diesel @ Rs.8.0/Litre	:	Rs. 2.0 lakhs
— Variable cost anticipated	:	Rs. 0.90/KWhr

list of developmental areas has been finalised. The R&D projects which are being developed include bagasse drying, bagasse compaction and cogeneration cycle optimisation, etc. Efforts are being continued to develop specific goal-oriented in each of these areas. Such projects are expected to be provided with a much stronger thrust during the Ninth Plan (1997-2002).

FINANCIAL SUPPORT THROUGH IREDA

IREDA continued to play a supportive role in the promotion of biomass power during the year. It has been sanctioned a US $ 100 million line of credit from the Asian Development Bank, of which US $ 20 million has been earmarked for financing cogeneration projects. The IREDA has approved Rs. 86.00 crore (860 million) as term loans for biomass power in nine private sector projects, involving an aggregate surplus power

TABLE 4.3: **Policy Announcements of Different States**

S.No.	Items	Maharashtra	Tamil-nadu	Karnataka	Uttar Pradesh	Madhya Pradesh	Punjab	A.P.	Haryana
1	Participation permitted (other than Govt. Sector)	Coop	Private	Private	Private	Private	Private	Private	Private
2	Rates/Charges for:								
	a) Power wheeling	2% of energy generated	15% of energy generated 2% to sister concern within 25km	6% of energy generated	12.5% of energy generated	2% of energy generated	2% of energy generated	15% of energy generated	10% of energy generated
	b) Power banking								

capacity of about 63 MW. Two projects having a surplus power capacity of 14 MW were commissioned.

STATE GOVERNMENT POLICIES

The Ministry has continued its efforts to persuade the State Electricity Boards/State Governments to announce remunerative policies for purchase/wheeling/banking of power from biomass projects. During the year, Haryana and Andhra Pradesh announced their policies for power from biomass projects. The Government of Gujarat is also reported to be seriously considering announcement of general policies. A summary of the latest policy announcements in various States is given in Table 4.3.

CONCLUSION

Taking into account the experience on village electrification through biomass gasifiers in the past few years it may be a techno-economically viable option for village electrification in India. This option has also large option for employment generation in rural areas.

ANNEXURE-I

State-wise Unelectrified Villages as on 31.3.96 (CEA Data)

Sl.No.	State	Total no. of Villages (91 Census)	Villages Electrified	Villages Unelectrified
1.	Arunachal Pradesh	3649	2270	1379
2.	Assam	24685	21887	2798
3.	Bihar	67513	47805	19708
4.	Jammu and Kashmir	6477	6274	203
5.	Madhya Pradesh	71526	67741	3785
6.	Manipur	2182	2016	166
7.	Meghalaya	5484	2407	3077
8.	Mizoram	698	626	72
9.	Orissa	4698	33871	13118
10.	Rajasthan	37889	30620	7269
11.	Tripura ($)	855	758	97
12.	Uttar Pradesh	112803	86639	26164
13.	West Bengal	37910	29205	8705
	Total (States)	418660	332119	86541*

* Excludes depopulated and uninhabited villages and States/UTs which have already achieved 100% electrification level.

$ -1981 Census

5

BIOMASS
A Reality for Cleaner World

INTRODUCTION

The growing population in developing countries has created unprecedented challenges to the humanity. Corresponding requirements of natural resources to fulfil the needs of the masses in the developing world is formidable. Social development is no more considered a responsibility of the developed world (as was the case in the past). Ever since the World Summit on Social Development held at Copenhagen in 1995, it has become apparently clear that the developed world is no more willing to extend help at the levels that were available earlier. A global call for maintaining aid to developing world at 0.7% of the GNP was totally negated. The present level of such aid amounts to 0.25% of the developed world's GNP and is likely to decrease as time passes. It does not mean that the two way linkages between the developed and the developing world are going to cease, but it is based more on barter deals and on economic issues. Correspondingly, the future fund transfers will be on commercial basis through the private channels.

Paper presented at the International Symposium on "Advances in Alternative/Renewable Energy Technology and Environment Issues", organised by University Technologi Malaysia at Johor, Malaysia during 22-24 July, 1997.

Energy has played an important role in the socio-economic development. World population is expected to double by the middle of the 21st century, and economic development will continue at a faster pace in the developing world than in the developed world. The likely result is a 3-5 fold increase in the world economic output by 2050 and a 10-15 fold increase by 2001. By 2100, per capita income in most of the currently developing countries will have reached and surpassed levels characteristic of the developed countries today, making current distinctions between the two totally obsolete. Despite rapid economic development, it is not necessary that adequate energy services will become available to every citizen in the next hundred years. Demand for energy services will grow, as much as an order of magnitude, 2050. Primary energy demands will grow less due to improvements in energy intensities. The focus, therefore, will shift from primary energy to the final energy form. Future will see emergence of new generation of energy utilization equipment which will be attaining the threshold efficiency limits of energy use. It is expected that the primary energy requirement will increase by 1-1/2 to 3 fold by the year 2050.

Resource availability is not being considered a constraint to the year 2050, but its geographic location is a major concern. The economic development over the next century, therefore, may not be constrained due to lack of energy resource. At the same time, one cannot rule out regional shortages and price increases due to unequal distribution of fossil resources. The law of averages cannot be uniformly applied to all countries, indicating that they are sufficient in energy supplies. Environmental concerns, financing and technology needs appear a more likely source of future limits; and international politics may also intrude from time to time.

According to WEC Study, energy services will move towards more flexible, convenient and clean fuels reaching the consumer. High quality energy forms become transitional fuels of choice. Thus, there will be an obvious shift towards electricity and towards higher quality fuels, such as natural gas, oil products, methanol and eventually hydrogen. Final energy structures appear to indicate similar patterns, while the primary energy requirements would diverge. That, in other words, means that the final forms of energy requirements for socio-economic development under different scenarios of development indicate the requirements of similar forms of energy, but when it comes to supply side primary energy sources, the different input sources are expected to double. No matter what the eventual dominant fuel in a scenario, there is a shift away from non-commercial and mostly unsustainable uses of biomass, and direct uses of coal virtually disappear. That obviously leads to another important aspect and that shows that biomass use will not be curtailed or stopped, but it will shift away from the traditional biomass use to its conversion through high efficiency modern combustion technologies to higher forms of energy supply like electricity, etc. (at the highest possible efficiency). Fossil sources continue to provide world's energy well into the next century, but in varying degree. Sustainable uses of renewables come to hold a prominent place in all scenarios reaching the consumers as electricity, liquid or natural gas, rather than solids in the present case of biomass.

Environmental considerations will be holding the key to the future energy consumption patterns. The natural capacity of the environment to absorb higher levels of pollution, particularly in densely populated metropolitan areas, will become the limiting factor for the use of fossil fuels. Another urgent environmental problem is the high concentration of particulate matter and sulphur dioxide in urban areas. Here, both cleaner fuels and active abatement measures will be required. Regional air pollution could prove problematic, especially in the rapidly growing densely populated coal intensive economies of Asia.

The continuing shift to higher quality fuels means continued decarbonisation of the energy system, and lower adverse environmental impacts per unit of energy consumed. Additional active policies will be required to reduce global carbon dioxide emissions. WEC expects that rapid technological improvements mean quicker progress towards clean fuels desired by consumers, and cleaner fuels mean a cleaner environment. It is in this context that the biomass technologies have an edge over any other technology.

BIOMASS AS A SOURCE OF ENERGY

Biomass, in all forms, supplies about 14 per cent of the world's primary energy. In the developing world, fuelwood provides a source of energy for basic sustenance, i.e., for cooking; and in the developed world, fuelwood is used either for power generation or process heat. In view of specific nature of fuelwood, it is presently the largest energy source in the developing countries where it produces 35 per cent of the primary energy (90 per cent in some areas). Since nine-tenths of the world population may reside in the developing countries, but the middle of the 21st century, biomass energy is likely to remain an important global energy source. The 15th Congress of the World Energy Conference held at Madrid in September 1992, emphasized this aspect and was reiterated at the 16th Congress held at Tokyo in October, 1995.

The World Bank, in its recent report : "Rural Energy and Development", 1996 has referred to the "nearly two billion people in developing countries (whose) search for energy is a daily grind". They have neither electricity nor gas nor oil to cook their food. The World Bank notes :

> "Women and children from these families often spend hours each day gathering dung and wood—hours they might otherwise be able to spend on producing work or education. Their health is damaged by the smoke given off by these fuels. Their environment may be damaged

as they chop down trees for wood."

The World Bank also notes that in South Africa, traditional biomass accounts for about 55% of total rural energy use, not much more than 25% of rural population, and 50% of urban population are connected to electricity.

The solar initiative of the World Bank embodies the work with its member countries, research and development bodies, industry representatives and NGOs to hasten commercialization of renewable energy technologies and to expand their applications in developing countries.

The Committee on New Renewable Sources of Energy and on Energy for Development of the U.N. released a policy statement on February 17, 1995 which analysed rural energy needs and prospects in developing countries on the basis of scenarios indicating that the population in the developing world is expected to increase to over 7 billion in 2025, with rural population remaining around 3 billion (as most of the increases in the developing world are to take place in urban centres). Per capita income and productivity in rural areas remain low and poverty is rampant. Income from agriculture is only US $ 300 per capita. Numerous and often grave institutional shortcomings and the lack of basic infrastructure such as energy and water supplies, transport and communications, all contribute to the problems faced by rural areas.

In Africa, Asia and Latin America, 80% of total rural energy consumption is absorbed by households, mainly for the preparation of food; 15% is used for agriculture, trade and industry; and 5% is consumed in transportation. Rural areas in these countries are, for the most part, dependent on draft animal power and traditional fuels such as fuelwood, agricultural and animal wastes. More modern technologies are also playing increasingly important role in some of the developing countries. Coal, crude oil, furnace fuel oil, natural gas and hydropower are used for electricity. It is estimated that during the period 1970-1990, developing countries provided electricity to an additional 1.25 billion in both urban and rural areas. Despite this effort, out of about 4 billion people in the developing world, 2 billion, mostly in rural areas, are still without electricity.

The Committee felt that the challenge now is to develop a strategy to facilitate and expedite the transition towards the use of efficient agro-technologies as well as an appropriate mix of traditional, conventional and renewable sources of energy, in rural communities and economic activities presenting a variety of socio-economic settings. Such a strategy would consist of energy options which would include:

(i) renewable energy sources such as photovoltaic systems, solar thermal energy, biomass energy and hydro power;
(ii) fossil fuels like coal, oil and natural gas.

PRESENT PATTERN OF GLOBAL ENERGY USE

Over the past 30 years, the world's energy requirement has risen considerably. In 1960, the world used 3.3 Gtoe (giga tonnes of oil equivalent) of energy; and in 1990, 8.7 Gtoe, an increase of 163 per cent or an average increase of 3.3 per cent per annum.

Energy use per capita is even more varied. At present, the majority of the world's population uses less than 1.2 toe (tonnes of oil equivalent) of energy per capita (indeed many use less than a third), the energy rich (in North America) use more than 7 toe per head. Under the commercial energy, the per capita energy consumption in India is still one of the lowest in the world. It recorded 123.5 kgoe (kilogrammes of oil equivalent) during 1990-91. However, if non-commercial energy consumption is also included, the per capita consumption goes up to 265.4 kgoe. The per capita electricity consumption in India recorded 253 kwh. However, the growth in demand has been fastest in the developing countries. In the past decade, it has risen in developing countries by 49 per cent, and in the developed countries by only 14 per cent. Although most of the energy is still consumed by the developed industrialised countries, the proportion accounted for by the developing countries can be expected to continue to increase.

GLOBAL ENERGY PERSPECTIVES—WEC REPORT: "ENERGY FOR TOMORROW'S WORLD"

The WEC Commission has developed three Global Energy Cases, each representing a different assumption in terms of economic development, energy efficiency and environmental impact.

(i) The first case 'A' is referred as the Reference Case (REF). It is in essence the same as developed by the WEC for the Montreal Congress in 1989, updated by recent experience, and considering "business-as-usual" scenario';

(ii) The second case 'B' is the Enhanced Economic Development Case (EED) which assumed a somewhat higher rate of economic growth in the developing countries than the Reference Case but with somewhat reduced energy efficiency; and

(iii) The third case 'C' is of the Ecologically Driven Case (ED) which assumes the same economic growth as in the Reference Case, but with a sharper improvement in energy efficiency.

WEC in its report "Global Energy Perspectives to 2050 and Beyond", published in 1995 further refined these cases into six different scenarios.

Case 'A' includes three scenarios addressing key developments in energy supply. They vary principally and that in the future they envision coal on one side, and nuclear and renewables on the other. In the A1 scenario, there is a large future availability of oil and gas resources. Dominance of oil and gas is perpetuated to the end of the 21st century. At the other end of the spectrum, the A2 scanario assumes oil and gas resources as scarce, resulting in a massive return to coal. Finally, in the A3 scenario, rapid technological change in nuclear and renewable energy technologies results in a phase out of fossil fuels due to economic reasons rather than resource scarcity.

Case 'B' has only one scenario and reflects a pragmatic approach. It manages to fulfil the development aspirations of the South, but less uniformly, and at a slower pace. For regions such as Africa, progress is painfully slow. Overall, economic growth rates to 2020 are more modest than in Case 'B' of Energy for Tomorrow's World.

In Case 'C' the nuclear energy is at crossroads. Two scenarios are included. They both meet the Carbon Dioxide emission ceiling in 2020, but they describe two very different paths that nuclear power might take.

Future energy supply in the three cases in the foreseeable future is projected to be as follows :

TABLE 5.1: **Future Energy Demand — Three Cases (Gigatonnes of Oil Equivalent, Gtoe)**

1990	2020		
	REF	EED	ED
8.7	13.3	17.2	11.2

Table 5.2 reflects the possible "Global Energy Supply Mix" for the three cases in the year 2020 as compared to the corresponding energy mix in 1990 :

TABLE 5.2: **Energy Mix in Global Fuel Use (Gigatonnes of Oil Equivalent, Gtoe)**

Energy Form	1990	2020		
		REF	EED	ED
Coal	2.3	3.2	4.8	2.1
Oil	2.8	3.7	4.6	2.7
Gas	1.7	2.8	3.5	2.3
Nuclear	0.4	0.8	1.0	0.7
Hydro	0.5	1.0	1.2	0.9
New Renewable	0.2	0.5	0.8	1.5
Traditional* Biomass	0.8	1.3	1.2	1.0
Total	8.7	13.3	17.2@	11.2

* Only firewood, agro residues and cow dung and does not include biomass recoverable by new technologies.

@ Rounding of the figures show a deviation of 0.1 in the total.

As per the WEC Study of 1995, energy supply in 2050 under different scenarios is expected as given in Table 5.3:

TABLE 5.3: **World Energy Supply Mix in 2050 (Gtoe)**

	A1	A2	A3	B	C1	C2
Coal	3.8	7.8	2.2	4.1	1.5	1.5
Oil	7.9	4.8	4.3	4.0	2.7	2.6
Gas	4.7	5.4	7.9	4.5	3.9	3.3
Nuclear	2.9	1.1	2.8	2.7	0.5	1.8
Renewables	4.7	4.9	6.8	3.7	4.8	4.2
Non-Comm.	0.8	0.8	0.5	0.7	0.8	0.8
Total	24.8	24.5	24.5	19.7	14.2	14.2

Present developing country energy supply mix in 2050 is

given in Table 5.4 :

TABLE 5.4: **Present Developing Countries Supply Mix in 2050 (Gtoe)**

	A1	A2	A3	B	C1	C2
Coal	2.1	4.8	2.0	2.8	1.4	1.4
Oil	4.7	2.7	2.6	2.5	1.8	1.8
Gas	2.6	2.4	3.5	2.1	1.9	1.8
Nuclear	0.9	0.5	1.08	1.0	0.1	0.7
Renewables	3.4	3.3	4.7	2.8	3.7	3.1
Non-Comm.	0.7	0.7	0.4	0.6	0.7	0.7
Total	14.4	14.4	14.2	11.8	9.6	9.6

Renewable energy availability in 2050 is given in Table 5.5:

TABLE 5.5: **Renewable Energy Availability in 2050 (Gtoe)**

	A1	A2	A3	B	C1	C2
Modern Biomass	1.1	2.2	3.2	1.4	1.6	1.5
Solar	1.8	0.4	1.6	0.4	1.3	1.4
Hydro	1.0	1.1	1.1	0.9	1.0	1.0
Wind	0.7	0.6	0.8	0.6	0.6	0.5
Other	0.1	0.6	0.1	0.4	0.3	0.8
Total	4.7	4.9	6.8	3.7	4.8	5.2

PROJECTION OF GLOBAL CONTRIBUTION OF BIOMASS

Biomass provides the possibility of conversion into liquid fuels, solid fuels and gaseous fuels. This versatility opens new opportunities for its use for different applications. This will make it significantly important for use in the 21st century. It will also continue to meet its important traditional roles, i.e., cooking, heat, light, construction, etc. in the rural and urban sectors of the developing countries. Biomass is a flexible feed stock capable of conversion into an appropriate form which can substitute for fossil based fuels at a relatively small scale investments. Many benefits may additionally accrue from significantly increasing the use of modern forms of biomass energy. These include the reduction of carbon dioxide emissions into the atmosphere if advanced biomass conversion technologies are used to substitute for fossil fuel-based technologies. Further more, the adoption of advanced biomass energy conversion systems may well be less costly than equivalent advanced fossil fuel-based systems.

WEC has estimated biomass energy use potential in 2020 at 1,566 Gtoe under the current policies scenario; and 1,711 Gtoe under the ecologically driven scenario. J. Wood and D. Hall in their study "Bio-energy for Development"—FAO Environment and Energy Paper 13 (1994) have estimated that currently biomass provides about 14 per cent of the global primary energy consumption and could in the future, theoretically, supply the equivalent of about thrice the present global energy demand. This scenario allows for increases in the crop land area for food production by the year 2025. More recently, the perception of biomass energy has changed in a number of industrialized countries. This has led to biomass gaining a growing and significant share of the primary energy sector in the USA, Sweden and Austria (4, 16 and 10 per cent of the primary energy respectively). Biomass is increasingly viewed as an environmentally and socially advantageous source of energy.

Biomass could become a central part of future sustainable energy supply, through both economic and technical feasibility. Such a developmental approach has been demonstrated by Johannson *et al.*, (1992) in their "Renewable Intensive Global Energy Scenario (RIGES)". It is demonstrated that there is a possibility of providing energy for growth and development at no extra cost compared to conventional fossil-based systems. A reduction in the global carbon dioxide emissions would occur as a result of such an increase in biomass-based energy. Under RIGES, by 2050, biomass could provide 17 per cent of electrical power and 38 per cent of direct fuel use. Altogether, "Renewable" could supply three-fifths of electrical power production and two-fifths of the direct fuel use, by 2050, at the same or lower cost of fossil fuel-based systems.

Biomass use has been finding better places as technical advances are making biomass a more promising energy source, and taking into account the environmental costs of power generation which all works in its favour. The most effective power stations using biomass at the moment are "combined cycle" plants. These can turn more than 50 per cent of the energy in a gaseous fuel into electricity, compared with only 35 per cent in the best coal-fired plants. A French Study (1992) has drawn attention to the employment implications : producing energy from biomass, it is reckoned, is three to six times as labour intensive as using fossil fuels.

D. Anderson in Report on "Renewable Energy Technologies" (1992) for the World Bank has emphasized three areas where policies that would begin to address the problem of global warming would also be beneficial to the developing countries. The first is the development of renewable energy sources wherein well-researched possibilities in the biomass and other sources of renewable energy exist. He has strongly recommended that the biomass technologies along with solar technologies have an immense promise, and may very well offer some economic and commercial surprise. In another study of the World Bank on "Renewable Energy Technology—A Review of Status and Cost" (1993) emphasis has been placed on three sources, namely, photoelectric, solar thermal and biomass for the production of electricity and liquid fuels.

Thomas B. Johannson *et al.*, in their book, "Renewables

for Fuels and Electricity" (1992) have projected that the liquid and gaseous fuels derived from renewable energy sources for transport and other applications would be marketed much like oil and natural gas are today. Biomass grown in a sustainable manner would play a major role in both developing and industrialized countries. Biomass would be converted into electricity and liquid and gaseous fuels using modern conversion equipment that is far more energy efficient than the biomass conversion equipment now being used.

Robert H. Williams and Eric D. Larson in their paper, "Advanced Gasification Based Biomass Power Generation" published in "Renewables for Fuels and Electricity" (1992) has reflected that a promising strategy for modernizing bio-energy is the production of electricity or the co-generation of electricity and heat using gasified biomass with advanced conversion technologies.

As regards the biomass production, productivity values of plantation have been reported to be upto 40 tonnes per hectare for high intensity plantation. However, continuous production facilities at 10-15 tonnes per hectare per year may be quite satisfactory. David Hall has suggested that countries like India should plan their long term policy of biomass plantation on an average yield level of 4 tonnes per hectare per year only.

The high yields achieved by intensive agriculture presently require 'energy intensive' and 'mechanized methods' of production. For certain crops under intensive management, the energy required for cultivation and processing may exceed the energy content of the food produced, excluding the energy in residue. However, the energy output to input ratio of woody and fibrous energy crops is very favourable (in excess of ten times and about six times for ethanol from sugarcane). In general, cereal crops give a positive energy return even under intensive management, e.g., maize contains 3.5 times energy in the harvested grain than it required to cultivate and process. Some crops presently require more energy to produce than they return in the food produced, e.g. energy output to input ratios are : apples 0.9, lettuce 0.2, tomatoes 0.6 and cabbage 0.8. However, these calculations do not account for the energy content of the associated residues which are significant. These figures are based on the recommendations of Pimentels, as quoted by J. Wood and D. Hall in 'Bio-energy for Development'—FAO Environment and Energy Paper 13 (1994).

The biomass provides an excellent opportunity to make farms energy exporters and net carbon sequesterers (sequester is defined as the net remover of carbon dioxide from the atmosphere and subsequent storage in organic matter).

POTENTIAL GLOBAL PRODUCTIVITIES OF BIOMASS

Total crop land increases have been at the expense of forests and woodland, with arable land estimated to have occupied 860 mha in 1982 and 1,477 mha in 1989. At the same time, forested land decreased from 5,200 mha to 4,087 mha (31.2 per cent of the world's land area).

The production of agricultural surplus in Europe and North America has attracted large subsidies for lowering the production. Wright L.L., *et al.*, in 'Opportunities to Mitigate Carbon Dioxide Build Up Using Short-Rotation Woody Crops' in "Forests and Global Warming", Vol.-I of American Forestry Association (1992) estimate that payment for the modern farm programmes in the USA is costing "one and a half times net farm income, whilst global agricultural subsidies were estimated to be about US $ 260 billion, in 1990. If present policies continue as usual, this will climb to US $ 300 billion by 2000".

It is estimated that in the United States, about 30 mha have been removed from crop production by 1988. In the European Community (12 countries) surplus agricultural land resulting from rising yields and changing agricultural subsidies may reach 15-20 mha by the year 2000, and at least

40 mha in the following centuries if the productivities continue to increase. These crop lands are already being removed from the intensive farming under set-aside schemes of the European Community and "Crop Land Reduction Scheme" of the USA. According to D. Hall, this land represents a significant opportunity to initiate biomass energy production schemes, especially if coupled with the use of agricultural and forestry residues.

Estimates of degraded and abandoned land generally lie between 700 and 1,000 mha which is equivalent to about half the world's present arable land. The extent of this available land has led the scientists to highlight its potential for use in mitigating the Green House effect by managing it to become a carbon sink. Wastelands are regarded as having a good potential for storing carbon in trees due to the relatively low levels of carbon in their soils and vegetation.

According to D. Hall, at good productivities (referring to 6t carbon/ha/year equivalent to about 12 oven dry t/ha/year biomass) the reforestation of all this land could theoretically remove about 5 GtC from the atmosphere per year over the next 40 years. Present emissions of carbon dioxide into the atmosphere are around 7-8 GtC/year of which 3-4 GtC appears as the atmospheric build-up in the levels of carbon dioxide. Sequestration programmes could, therefore, be regarded only as a temporary measure, buying time until other sustainable forms of energy or permanent carbon dioxide removal systems can be developed. It would require large quantities of inputs in the form of management, fertilizers, pesticides and labour in order to raise the productivity significantly.

Majority of degraded soils are salt affected. High salt levels reduce the levels of nitrogen which is available to the plant, but nitrogen fixing species are often tolerant to saline soils, and may achieve acceptable productivities on such land. The potential for use of saline land is considerable, for example, salt tolerant plants can attain 3 to 7 tC/ha/year with saline water irrigation.

ROLE OF ENERGY PLANTATION IN CARBON SEQUESTRATION

Sustainably grown biomass is nearly carbon dioxide neutral, depending on production and conversion methods. If biofuel production is to be regarded as a carbon sink, the total amount of carbon stored per hectare under the bioenergy crop must be greater than the level of stored C in the vegetation previously on that land. Also where biofuels are used as substitutes for fossil fuels, the biofuel can be regarded as a sink in terms of the avoided carbon dioxide emission that would have been derived from fossil fuels.

The most clear benefit in carbon sequestration terms would be if the plantation was established in a desert with an existing standing stock of almost '0' tC/ha. Both the above and below-ground levels of carbon would be raised significantly compared to the existing level. However, if the standing stock of the previous vegetation was greater than that of the new plantation, then a net reduction in the standing C levels would result. If the plantation biomass were to be used as a fossil fuel substitute (or long-lived produce), a new reduction to atmospheric carbon dioxide emissions would occur after a given time period. The high the levels of vegetation, prior to the plantation, the longer it would take for fuel substitution benefits to reduce the amount of carbon dioxide initially released. This results from the relatively large amounts of carbon dioxide which are released during the initial clearance through harvesting and transport losses.

Marland G. and Marland S. (1992) in 'Should We Store Carbon in Trees' have explored the use of plantations for a fossil fuel substitution plantation-based model (1992). They conclude that the three most important factors in assessing whether biomass energy plantations are effective carbon sinks are as follows :

(i) The carbon inventory of the natural vegetation;
(ii) The productivity of the plantation; and
(iii) The time perspective adopted.

Biofuels provide an opportunity to avoid fossil-fuel-derived carbon dioxide emissions. However, the present intensive agricultural methods can be used only for the production of liquid fuels, e.g., diesel from rape seed, if the associated residue production is used for energy. Thus, the use of carbon-flows and energy output : input ratio to monitor net carbon dioxide production must be integral to the strategies for energy provision.

ROLE OF NEW CONVERSION TECHNOLOGIES IN PROMOTION OF BIOFUELS

Following approach in biomass conversion technologies has been advised :

(i) The "traditional" domestic use in the developing countries for household cooking, lighting and space heating. In this role, the efficiency of conversion of the biomass to useful energy has been raised from 5 per cent to 20 per cent.
(ii) The "traditional industrial" use of biomass for the processing of tobacco, tea, pig iron, bricks and tiles, etc., where the biomass feed stock is often regarded as a free energy source. There is generally little incentive to use the biomass efficiently and so the conversion of the feed stock to useful energy commonly occurs at an efficiency of 15 to 30 per cent.
(iii) The "modern industrial" activities relate to thermal conversion technologies. Expected conversion efficiencies are between 30 and 55 per cent.
(iv) Newer "chemical conversion" technologies (fuel cell) are capable of bypassing the entropy-dictated Carnot

cycle limit which describes the maximum theoretical conversion efficiencies of thermal units.

(v) "Biological conversion" techniques, including anaerobic digestion of biogas production and fermentation of alcohol.

CONCLUSION

WEC has identified important role of biomass in making energy cleaner than at present. Such a situation needs commercial technology development and government policies to go hand-in-hand. Whereas biomass growth is a good approach for carbon sequestration, but that cannot be an ultimate solution. A move towards efficient use of fossil fuels and higher use of biomass only can keep the environment clean. Theoretically, the biomass growth can substantially reduce pollution due to energy use. No doubt, the developing countries emit low carbon dioxide emissions per capita and have not bothered much for pollution controlling technology. The developed countries have developed a number of technologies. These technologies, however, cannot be transferred off-the-shelf from developed to developing countries. Every technology has to be site-specific. Therefore, biomass growth in the developing countries provides a good option.

6

PROMOTING BIOMASS ENERGY UTILIZATION THROUGH VILLAGE LEVEL SELF-GOVERNANCE INSTITUTIONS

INTRODUCTION

Biomass energy is the local energy available for meeting the minimum rural needs of cooking. Though the contribution of biomass sourcess in the overall energy scenario is gradually decreasing but it still contributes over 40% of the energy supply in the country. 65% of the biomass energy in the rural areas is apportioned to fuelwood, 20% to agricultural waste and 15% to cow dung. With the increasing use of commercial sources of energy there has been a substantial shift towards commercial sources recently. As such the future projections for energy in India do not show a proportionate increase in the fuelwood consumption with the rising population. It is difficult at this stage to predict the shift in the fuel-mix but it is clear that shift is taking place. Also, in view of the global energy policy considerations the final form of energy is more important than the primary form of energy. Therefore, there has been a major thrust on how the fuelwood and other sources of energy can be converted into desirable form, thereby making the primary sources of energy of secondary

---

Paper presented at the XI World Forestry Congress, Technical Division-16, at Antalya, Turkey during October 13-22, 1997.

importance. This change is gradual but quite perceptible.

The Government of India has focused attention on governance at the rural level through Panchayats, the body of elected members of the public. The Panchayats have also been assigned certain development tasks as laid down in the 11th Schedule of the Constitution of India. Social forestry and farm forestry; along with land improvement, implementation of land reforms, land consolidation and soil conservation; fuel and fodder; and non-conventional energy sources are the responsibility of the Panchayats.

It is necessary to have a look at the energy policy and rural energy planning efforts made in India, the ongoing programmes to consider how the biomass production can be better managed and regularised through the local governance systems.

ENERGY PLANNING

India has considered short term, medium term and long term energy planning process in the country.[1]

In the short term, the effort is to maximise returns from the assets already created in the energy sector, improving efficiency in production, transmission and end use; reducing energy intensity of different consuming sectors and initiating steps for meeting fully the basic energy needs in the urban and rural households.

In the medium term, progressive substitution of petroleum products by coal, natural gas and electricity; accelerated development of renewables and promotion of R&D efforts on decentralised energy technologies based on renewable resources have been suggested.

In the long term, promotion of energy supply system based largely on renewables and promotion of technologies of production, transportation and end use of energy, that are environmentally benign and cost effective have been suggested.

RURAL ENERGY PLANNING

Though fuelwood, agro-residue and cow dung are the main fuel sources, only use and availability of fuelwood can be planned and quantified. Primary data is available on fuelwood from field surveys, conducted by the National Sample Survey, Organisation (NSSO), National Council of Applied Economic Research (NCAER) and other Research Institutions and individuals.

J.N. Maggo, in his presentation to the Fuelwood Planning Course of RWEDP in 1995 summed up India's efforts at rural energy planning as follows:[2]

The methodology of estimation of household energy demand was first given by the Energy Survey of India Committee (ESIC) in 1965. Based on the results of the sample survey carried out by the agencies mentioned above, the ESIC adopted a norm of 0.39 tonnes of coal replacement (tcr) of energy consumption requirement per person per year in the urban households and 0.38 tonnes of coal replacement of per capita energy consumption per annum in the rural areas. After so determining the total energy requirement in the household sector, the projected contribution of commercial sources of energy was subtracted from it to arrive at the share of non-commercial fuels. The same was then apportioned among fuelwood, agricultural waste and cow dung in the proportion of 65:20:15. Later in 1986 the Energy Demand Screening Group (EDSG) set up by the Planning Commission modified the consumption requirement of useful energy to 520 kilocalories per person per day in the rural areas. Other studies carried out later like the Sectoral Energy Demand in India also estimated the useful energy requirement in 2009-10 as 641 kilocalories per person per day.

In a number of studies conducted by IAAS, it was observed that the rural domestic energy consumption in the final form varied from 420 kilocalories to 620 kilocalories per capita per day, depending on the status of the family and the use whether only cooking or cooking and water heating were both required.

The ABE Report 'Towards a Perspective on Energy Demand and Supply in India in 2004-05' projected that atleast 60 million hectares of land (comprising open/degraded forest land, barren and unculturable land, culturable wasteland, land other than current fallows and farm perimeter land) might be available for raising fuelwood/fodder plantations. The study brought out that the productivity of the Indian forest land is quite low and need to be increased.

FIELD PRODUCTION OF BIOENERGY CROPS

Subsequent policy decisions to separate community based/ farm forestry energy plantations from large forest lands has been found a good option. Various studies in places like Haryana, Kerala and Western Uttar Pradesh have indicated that the local availability of fuelwood from village woodlots has resulted in stabilising the local market price of fuelwood.[3]

Such an effort has been possible only after sustained steps taken by the Department of Forests, the Wastelands Development Board and the Ministry of Non-Conventional Sources for promoting Research and Development for evolving new varieties of fuelwood which can be grown under different conditions with high yield and short rotation basis.

Different concepts so far introduced in India for improving the growth of biomass for energy have been the following:

Forest Management

Under the Afforestation Programme, emphasis has always been on timber growing. Planned development started in the year 1951 and substantial improvement in forest preservation started during the Seventh Five Year Plan (during 1985-1990). Progress under afforestation during 1951-1992 has added 18.835 million hectares of area under vegetation. This programme is totally under the government forestry department.

Social Forestry

Under the Social Forestry Programme, the main objective is to produce fuelwood, fodder, small timber and other products to meet the needs of rural communities by raising plantations on wastelands including community lands, degraded forests, other government wastelands and private marginal lands. Under government projects, funds were given only to government or aided institutions. However, it was observed that most of the degraded lands were under private control and, therefore, the progress was slow.

Agro-Forestry

The third concept used for raising fuelwood through community participation has been under the agro-forestry programme, where farmers inter-crop cereal crops with tree plants on bunds, periphery of farmland on village land. This is mostly in an individual activity and very often the farmer does not find it financially lucrative.

Wasteland Afforestation

Under afforestation for reclaiming wastelands, a massive programme was developed by the National Wastelands Development Board with people's participation. The aims include the following:

(i) Establishment of decentralized nurseries and school nurseries,
(ii) Block Plantation especially on community land and lands owned by the scheduled castes and scheduled

tribes and people living below the poverty line; and

(iii) Pasture development through people's institutions and involvement.

Though 92 mha are reported available, only about 20 mha can be economically exploited. Expert Consultations, including those under FAO and IAAS, have indicated that energy plantations with long term yields of 4 to 5 tonnes per ha per year can be certainly grown on various wastelands and village lands.[4] Yields of upto 40 tonnes per ha have been recorded in test fields.

Biomass Research Centres

Biomass Research Centres under the Ministry of Non-Conventional Energy Sources function in nine different agro-climatic regions of the country and are involved in promoting 70 fast growing fuel wood tree species. The productivity of the selected species is assessed under field conditions and efforts are being made to develop detailed packages of practices, including standardisation of necessary requirements; application of bio-fertilizers; and insect, pest and diseases control measures. The BRCs have already finalised packages of practice for about 25 species. Efforts are being made to commercialise supply of quality planting material. Training programmes for different target groups including the members of Panchayats are being organised. Further, entrepreneurship development (local governance institutions) the field of nursery technology is being attempted.[5]

The BRCs have initiated field trials on about 25 other species to assess their performance in terms of fuelwood yield. The Lucknow Centre has initiated experiments on coppice rotation and high density plantations. Further, provenness trials and progeny testing of Prosopis juliflora, especially the thornless variety, have been taken up. The Bhubaneshwar Centre has propagated vegetatively elite male plants of

Casuarina equisetifolia with the aid of growth promoting harmones and is using them for field plantations. The Lucknow Centre has developed a low cost nursery technique to use water most efficiently and produce quality planting materials. All the BRCs are producing improved planting materials on a large scale and are supplying them to various user organisations for plantation purpose. About one lakh seedlings have already been marketed.

All the BRCs have been organising training programmes for different target groups, such as farmers, resource persons of implementing agencies, non-government organisations (NGOs), district and block level officers of various State Governments and representatives of industries and panchayats. Efforts have also been made to establish coordination with user organisations to supply elite planting materials. The Centres are also providing technical and consultancy services.

PANCHAYATI RAJ INSTITUTIONS

From ancient times villages in India had a tradition of self-management, the system of five (Paanch in Hindi) people conducting affairs of the village was known as Panchayat.

In 1978, a new generation panchayat system came up. The Ashok Mehta Committee recommended that Panchayats must be political institutions. West Bengal (1978), followed by Karnataka (1983) and later Andhra Pradesh, introduced this kind of Panchayat. During this period, because of the success of Panchayats in these states, the idea of a "District Government" also came up. A need was felt to go down to the block and village levels from the district level. The Indian Parliament in December 1992 passed the 73rd Constitutional Amendment and on 24 April, 1993 panchayats became Part IX of the Constitution.[6]

The striking features of the 73rd Constitution Amendment Act are:

i. Panchayats will be "institutions of self-Government".
ii. Gram Sabhas in villages comprising all the adult members registered as voters in the area will be the basic units of the Democratic system.
iii. Seats in Panchayats at all the three levels shall be filled by direct election and one-third seats will be filled by women.
iv. Specific responsibilities will be entrusted to the panchayats to prepare plans for economic development and social justice in respect of 29 subjects listed in the Eleventh Schedule. The 74th Amendment provides for a District Planning Committee to consolidate the plans prepared by the panchayats and municipalities.

The Panchayats are therefore, the institutions of self-governance and the Constitution gives them autonomy and powers to govern.

ROLE OF PANCHAYATS IN MICRO LEVEL PLANNING FOR FUELWOOD PRODUCTION

The role of Panchayats is important in raising biomass by utilising financial allocations under different schemes applicable at the village level.[7] Some of these will include the following:

i. Allocation under different plan schemes for programmes like, renewable energy sources, irrigation, soil conservation etc
ii. Allocation under integrated rural development programme and other employment generating schemes for the poor.
iii. Bankable schemes, that could be drawn on community basis or individual basis.
iv. People make own contribution on being convinced that investment made by them would mean creation of

assets in a cost effective manner. This will become possible with the proper interaction of the technical experts and scientists with the local people. The local government is expected to bring them together.

All these need appropriate micro-level plan. Provision has been made to give skills in such planning which can be best handled by the local governance institutions. These institutions can also play an important role in training for transfer of technology and its assimilation into day-to-day life. The expectation is that such a mechanism will lead to converting government schemes into productive assets for the rural areas to tap locally available resources for their sustenance. The BRCs have already launched programmes for training on fuelwood species through Panchayats.

India has 35.89 mha of degraded forest wastelands and 93.69 mha non-forest wastelands. Most of the non-forest wastelands is either under the control of local Panchayats (village level elected bodies) or individual ownership. Small farmers predominate over the large landowners. It is considered appropriate to launch a community based programme to achieve micro-level fuelwood availability. The emerging pattern of local self-governance can contribute significantly in achieving targets.

People would then, on their own be able to decide on the competing uses of land for biomass production for agriculture or for energy. A few examples that are already available have indicated that where alternate uses of fuelwood species are now coming up more and more people are growing fuelwood species. The micro-level availability through agro-forestry based on community's participation has already established self-sufficiency of fuelwood in a number of areas. This was very often ignored when a macro view of fuelwood was taken. The farmer now has a choice to select only such processes that are economically viable and financially beneficial to him and match it with the corresponding government aid.

CONCLUSION

Social forestry, farm forestry and renewable energy sources being important elements in the activities allocated to Panchayats, the village level institutions, certainly need allocation of larger funds to promote local level fuelwood production. In due course of time the village level institutions will also play an important role in developing and disseminating conversion technologies so that fuelwood supplies larger useful energy in the final form.

REFERENCES

1. Eighth Five Year Plan (1992-97)—Government of India.
2. Fuelwood Planning in India, J.N. Maggo—RWEDP Course on Fuelwood Planning—1995.
3. Biomass Production for Energy in India—David Hall-FAO-IAAS National Workshop on Bio Energy—1992.
4. Survey of Fuelwood Production under Social Forestry—Narayan Singh—Bio Energy Society of India Convention—1995
5. Annual Report of the Ministry of Non-Conventional Energy Sources, Government of India—1995-96.
6. Panchayati Raj Structure and its Linkage to Development, George Mathew—FAO—IAAS National Workshop on Panchayati Raj and the Food Security in India—1996.
7. Panchayati Raj Institutions—The Emerging Scenario—K.P.A. Menon —FAO—IAAS National Workshop on Panchayati Raj and the Food Security—1996.

ANNEXURES

ANNEXURE I

REGIONAL STUDY ON WOOD ENERGY: TODAY AND TOMORROW

EXECUTIVE SUMMARY

Purpose of the Study

The Asia-Pacific Forestry Commission (APFCF), at its meeting at Yangon in January 1996, agreed to a proposal for a Forestry Sector Outlook Study for the Asia-Pacific Region. The Study was to include scenarios for wood energy demand and supply in the context of overall energy transitions in the region.

The present study on 'Wood Energy Today and Tomorrow' has been prepared by RWEDP to contribute to the Outlook Study. It mainly considers the 16 RWEDP member-countries in Asia, which are Bangladesh, Bhutan, Cambodia, China, India, Indonesia, Lao PDR, Malaysia, Maldives, Myanmar, Nepal, Pakistan, Philippines, Sri Lanka, Thailand and Vietnam. The study links with the work initiated by FAO's Wood and Non-Wood Products Utilisation Branch (FOPW) on Wood Energy today and for the Future, which addresses the present wood energy situation at national, regional and international levels and possible wood energy scenario's for traditional as well as modern uses.

The study is aimed at policy makers in the forestry sector, but also touches upon policy issues in other sectors, e.g. agriculture, rural development, energy and environment. As yet, sufficient wood energy data are not available to enable woodfuels to be included in common energy modelling and supply-demand balancing. RWEDP is in the process of remedying this by continuing to collect and validate data. The study addresses energy issues only, and builds on available data on related subjects like population, landuse, demand and supply of timber products, etc.

WOOD ENERGY TODAY

Role of Woodfuels

Woodfuels are an important commodity from forests and other tree production systems. Each year, the 16 RWEDP member-countries meet about 10,000 PJ of their total energy consumption by woodfuel (which averages to 30%,excluding China). It has been estimated that woodfuels represent an economic value of about 30 billion US $ per annum. The current financial value of total industrial roundwood production in the same countries is about US $ 25 billion per annum. In Asia, the consumption of woodfuels is increasing in absolute terms by about 1.6% each year, and this trend is likely to continue in the foreseeable future.

Sources of Woodfuels

Woodfuels consist of woody biomass, i.e. stems, branches, twigs, etc., and saw dust and other residues from logging and wood processing activities, as well as charcoal from these sources. The primary sources of wodfuels are both forest and non-forest land. Forest and other wooded land includes natural forests (including degraded forests), scrub lands, wood and timber plantation and woodlots. Non-forest land here includes agriculture land, agro-forestry systems, waste land, line trees, home gardens, etc. The ratio between woodfuels originating from forest and non-forest land is generally not known, but data from 7 RWEDP member-countries indicate that about 1/3 of the woodfuels originates from forest land, and about 2/3 from non-forest land. Typically, non-commercial sources of woodfuels are located within a 20 km radius from the end-users, and commercial sources within a 100 km radius from the market.

Secondary sources of woodfuels are residues from logging and wood processing industries, but also recycled wood from construction activities, packing crates, pallets, driftwood, etc. In some areas recycled wood supplies as much as 20% of total woodfuels.

The "Fuelwood Gap Theory" Rejected

In the 1970s and 1980s it was generally assumed that all woodfuels originate from public forest land. This lead to the 'fuelwood gap theory' from which it was concluded that non-sustainable yields were taken from forests to meet wood energy demands. The theory resulted in the assumption that woodfuels use was a root cause of deforestation. But now, ample evidence exists to prove that the theory is false and that, except for localised areas, woodfuels use is not a main cause of deforestation. The indications are that a major cause of deforestation is the ongoing conversion of forest land into other land uses, particularly agriculture. This is generally carried out by planned forest clearing or results from gradual processes of forest encroachment.

Consumption of Woodfuels

Woodfuels are consumed mainly by rural populations, though substantial amounts are also consumed in most towns and cities. The larger part of the consumption is accounted for by households, but numerous industries and services are based on woodfuels. In the domestic sector consumption is spread over lower, middle and higher income groups. Even in conditions of increasing urbanisation and fossil fuel penetration, the large majority of the households in Asia use wood or other biomass fuels. In all RWEDP countries total consumption is still increasing, and in several countries woodfuels consumption per capita is also increasing. In fact, consumption per capita is very site-specific and influenced by factors like climate, households size, availability and reliability of supply of the various fuels and their potential substitutes as well as their costs, the appliances required for utilisation, and culture and tradition.

Most consumers in the domestic as well as the industrial and service sectors still avail themselves of relatively simple and inefficient technologies for combustion. Efforts are being made to disseminate improved technologies. Where successful, the efforts result in improved quality of life, particularly for women, or improved viability of traditional industries and services. However, as yet no evidence exists that the introduction of more efficient conversion technologies would have led to reduced demand for woodfuels from any forest resource base.

Substitution

Apart from wood, agricultural land produces biomass residues, part of which is available as fuel on an environmentally sustainable basis. At present, the main biomass fuels are crop residues like bagasse, rice husks and straw, coconut husks and shells, palm oil kernel, shells and fibre. Wood and other biomass fuels (as well as animal dung used for fuel) can substitute for each other, though most consumers have a general preference of wood over other biomass. In terms of energy content per ha per annum, the sustainable production of biomass residues available for fuel from plantations and agricultural land is about 30% of the sustainable yield of woodfuels from natural forest land. Non-sustainable production of potential woodfuels due to deforestation aggregated for the 16 RWEDP member-countries is about equal to the present woodfuel consumption (with variations per country from 10% to 1,400%). In RWEDP member-countries woodfuel represents about half of all biomass fuel consumption in energy units.

Woodfuels can also be substituted by fossil fuels, but this is not observed as a major or general trend in RWEDP member-countries (with notable exceptions). Rather, the current overall accelerated use of fossil fuels in Asia is mainly due to additional productive and consumptive activities in the modern sector of Asia economies. Such use is largely in addition to the wood/biomass fuels, rather than a substitute for them. The widely used term 'fuel transition' is often misleading, because what is actually going on is better described as 'fuel complementation'.

Environmental Aspects

Adverse environmental impacts of woodlfuel use are due to unsustainable extraction from environmentally sensitive areas, which can lead to degradation of watershed and catchment areas, loss of biodiversity and habitat, etc. However, if the supply sources is properly managed, woodfuel can contribute positively to the local and global environment. Woodfuel is CO_2 neutral, provided the rate of harvest equals the rate of re-grown. When wood and other biomass resources are properly valued by local populations their sustainable use contributes to the economical management of the local environment.

Wood is very beneficial for the global environment. It can be estimated that the net effect of woodfuel use in RWEDP member-countries in 1994 implied a saving of about 278,000 kton CO_2 which otherwise would have been emitted into the global atmosphere. If LPG is the hypothetical replacement of woodfuel the CO_2 saved will increase to 349,000 kton in 2010. These figures can be translated into costs avoided for recapturing the CO_2 and amount to 14 billion US $ saved in 1994 by woodfuel use in Asia, and 17.5

billion US $ saved in 2010. If coal is the hypothetical replacement of woodfuel the figures would double.

Social Aspects

Woodfuel supply implies labour for growing, harvesting, processing, wholesaling, transporting and retailing the product. Per unit of energy, the labour involved in these woodfuel businesses is about 20 times larger than for kerosene. Woodfuel business is the main source of income for about 10% of rural households, and for about 40% of their cash earnings. In times of hardship, or when harvests are insufficient for subsistence, the opportunity to generate income in woodfuel business provides a safety-net for poor persons, many of whom are women.

On the demand side, woodfuels are in basic commodity serving the daily needs of some 2 billion people in RWEDP member-countries. However, access to the fuels is very skewed. In areas or times of scarcity, landless and unemployed people and low-wage earners suffer from high prices or the non-availability of woodfuels.

Data Availability

Because of the site-specific and dispersed production and consumption of woodfuels (which are partly non-monetized), it is extremely expensive and time consuming to collect reliable and systematic data on woodfuel supply and demand. International organisations do not avail themselves of such data, and neither do most national organisations. Data on the same country published by different sources are largely conflicting, e.g., by a factor 3 or more. This is true also for industrialised countries. Sometimes data published by the same national source are also conflicting. It is further noted that in energy balances published by national sources, data on the supply of woodfuels are usually worked back from stated consumption figures. Therefore, most supply figures do not provide independent sources of information.

Data published by FAO are derived from baseline estimates made before 1961. For the RWEDP member-countries, these estimates have been annually updated under the assumption that the population elasticity of consumption is 1,000. Occasional surveys under the World Bank/UNDP ESMAP programme in the early 1990s revealed discrepancies of more than 100% with the FAO data, (it should be noted that such an occasional survey in one country can cost several million US $). Most country-level data on woodfuel consumption is based on ad-hoc surveys, and often industrial woodfuel consumption is not covered.

Prices

Prices of woodfuels vary, depending on markets. Part of the market is still not monetized (in most places some 70%). Commercial markets are generally found in cities or towns, but also in villages, where fuelwood is traded. Local prices are largely determined by opportunity costs of labour and resource availability, which generally does not reflect the real economic (including environmental) costs. A typical price in RWEDP member-countries is 40 US $ per ton. Stumpage fees can be anything between 0 and 20% of retail prices.

A significant increase in the usual price can be due to local scarcity, which implies that more time or labour is required to bring the fuel from a distant source to the consumption centres. For the more well-to-do consumers, price increases are generally not a reason for switching to fossil fuels (convenience of fossil fuels is) or another fuel. For poor consumers a price increase can be a reason to resort to cheaper, lower-grade biomass fuels.

Anecdotal evidence indicates some correlation between international oil prices and local retail prices of woodfuel. However, in general prices of woodfuels remain more or less constant in real terms. A very small fraction of woodfuels is for export, which fetches a relatively high price.

Policies

Both producers and consumers usually perceive woodfuels to be a by-product of wood products. This applies even in areas where woodfuel plantation or village woodlots have been established. The perception clearly contrasts with the real economic value of woodfuels as compared to wood products. The reason for the under-valuation of woodfuels may be that the benefits are largely dispersed over time and over numerous small consumers, and that the fuels are partly non-monetized. This is in contrast to the situation of wood products, whether forest departments, logging and processing companies, as well as individuals benefit from concentrated cash-flows. In other words, there are few or no powerful stakeholders in woodfuel matters.

Policy makers in the forestry sector can facilitate the sustainable production and good use of woodfuels without undermining the conditions necessary for the production of wood products. Adequate policies also support environmental management and social development objectives. To this end a number of specific policy measures in the forestry sector are identified in the present document.

WOOD ENERGY TOMORROW

As the available data are either patchy or unvalidated, it is not possible to develop a reliable quantitative outlook on wood energy in Asia. However, important trends and qualitative aspects of anticipated wood energy developments until the year 2010 can be presented. Interpretation of the data collected under ESMAP has provided insights into the main mechanisms governing the supply and demand of wood energy. According to research conducted or commissioned by

RWEDP these mechanisms still apply, and are likely to remain valid up to 2010.

Trends in Demand

The main trend on the demand side is the overall increase of consumption of woodfuels in all RWEDP member-countries by about 1.6% a year. The trends are based on extrapolations of time series of best available data which, in principle, incorporate influences of trends in relevant factors. The extrapolations do not take into account possible new effects of factors like accelerated urbanization, changes in household size and incomes, or culture and tradition because such data are simply not available. However, it is believed that the net effect of the combination of factors may be limited.

Prices

As no overall price elasticities for woodfuel consumption in traditional markets are known, it is not possible to base future demand estimates on trends in prices. In fact, it is likely that the overall price of woodfuels in real terms will remain more or less constant for the coming 5-10 years. Other biomass fuels may increasingly compete in traditional markets when their quality/price ratio improves. For modern applications, accelerated penetration of biomass fuels can be anticipated if, and when, fossil fuel prices increase substantially.

Trends in Supply Potential

Major trends regarding the suply side relate to changes in landuse, which differ for each country. The general trend in landuse change is a decrease of natural forest areas of about 3.6 million ha each year (1.2%), and an increase of agricultural land of 5.7 million ha each year (0.6%), aggregated for the 16 RWEDP member-countries. The net result is an increase in total forest and agricultural land of 1.1 million ha each year, which can be attributed to changes in other land uses (wasteland, etc.) Combined with an average twofold productivity for all biomass fuels from plantations and agriculture land as compared to neutral land, this should lead to an increase in the sustainable availability of total potential biomass fuels by 2010. In addition, there will be a non-sustainable supply of potential woodfuels because of the on-going process of converting forest land into agriculture and other land uses and commercial logging.

Outlook for the Demand-Supply Balance

The overall trends in demand and potential supply of wood and other biomass fuels per country are presented in this document. The aggregated results for 16 RWEDP member-countries in Asia are presented in Tables 1 and 2.

When reading the tables it should be noted that aggregation over a wide region in Asia leads to hypothetical supply availability. In reality, fuelwood markets are extremely localised and fragmented. Still, some general observations can be made from the data presented.

From Table 1, it is observed that by 2010 the sustainable aggregated potential supply of woodfuels still outweighs aggregated consumption. This positive balance does not depend on the accuracy of the estimates made, or on the assumptions incorporated into the estimates. A reasonable margin of error would still produce the same results. Moreover, the assumptions on supply potential are on the conservative side.

Much more important than the potential supply as such is its geographical and social distribution, since consumers may not be able to use available resources due to physical, financial and social constraints. The same comment applies to the country balances which are presented in this document. For some countries Table 2 suggests gaps between the

TABLE 1: **Consumption and potential supply of biomass fuels aggregated for the 16 RWEDP member countries**

	1994			2010		
	Area 1000 ha	Mass kton	Energy PJ	Area 1000 ha	Mass kton	Energy PJ
CONSUMPTION						
Total woodfuels		645,895	9,688		811,548	12,173
POTENTIAL SUPPLY						
Sust.woodfuel from forest land	416,204	669,812	10,047	370,363	629,339	9,440
Sust.woodfuel from agricultural areas	876,933	601,407	9,021	971,062	692,088	10,381
Sust. woodfuel from other wooded lands	93,140	53,994	810	81,368	47,170	708
Waste woodfuels from deforestation	(4,253)	605,565	9,083	(3,114)	437,710	6,566
Total potentially available woodfuels	1,382,024	1,930,778	28,962	1,419,679	1,806,307	27,095
50% of crop processing	876,933	218,915	3,458	971,062	322,024	5,105
Total potentially available biomass fuels		2,149,693	32,420		2,128,331	32,200

TABLE 2: **Potential supply and estimated consumption of woodfuels in 1994 and 2010 in 16 RWEDP member-countries**

	1994		2010	
	Potential supply (kton)	Estimated consumption (kton)	Potential supply (kton)	Estimated consumption (kton)
Bangladesh	8,999	9,396	9,271	13,320
Bhutan	5,946	819	5,624	1,195
Cambodia	81,565	5,375	43,827	7,553
China	598,546	219,122	639,733	252,819
India	235,167	173,412	255,729	225,725
Indonesia	439,049	54,474	394,923	67,465
Laos	46,006	2,329	38,902	3,496
Malaysia	137,301	6,187	97,777	8,216
Maldives	34	80	41	123
Myanmar	129,935	23,058	106,930	31,183
Nepal	11,444	12,987	10,304	18,378
Pakistan	22,569	34,687	21,144	52,167
Philippines	89,267	23,051	71,171	30,329
Sri Lanka	8,963	5,681	9,044	6,769
Thailand	67,030	46,069	59,157	53,390
Vietnam	48,960	29,368	42,730	39,418
RWEDP	1,930,778	645,895	1,806,307	811,548

estimated woodfuel consumption and the potential supply already in 1994, which cannot be real. The reasons are probably overly conservative estimates regarding supply, as well as complementation of wood by other biomass fuels. However, it is observed that Bangladesh, Nepal and Pakistan face pressures on their overall wood resource base, which are likely to increase.

Table 1, further illustrates that any deforestation process generates a large additional (potential) supply of woodfuels which, however, is not required for a positive balance of potential supply over consumption. If deforestation is due to conversion of forest land into agricultural land, the process would result in a sustainable increased supply of potential biomass fuels, because generally agricultural land has a higher biomass fuel productivity than forest land.

It is also observed that at present the sustainable potential supply of woodfuels from agricultural lands more or less can meet the consumption. The same applies for the aggregated sustainable potential supply from forest lands. However, the latter are more likely to be found in remote areas, whereas the former are generally closer to the rural consumers. This may explain the fact that most woodfuels originate from non-forest land, as shown by data for several countries. When looking at the total potential supply of biomass from agricultural lands, i.e. wood and crops residues together, it is observed that this can meet both the present and projected consumption.

From Table 1, it is further observed that, overall, the potential supply of biomass fuels in the form of crop-processing residues is substantial. It should be noted that the estimate builds on (only) half of the processing residues, leaving all field residues (which are about 4 times the processing residues) untouched. Indeed, local shift from wood to other biomass fuels can be anticipated to increase. This implies an immediate and increasing need for further development of cost-effective technologies for upgrading and combusting traditional fuels from crop residues, and for disseminating such technologies, and corresponding managerial systems. In the longer term there may be scope for expanding modern bioenergy fuels based on advanced R&D.

Supply Policies

While the overall supply/demand balancing of wood and other biomass fuels in the region will not be a major concern, their distribution will, as in the 1990s, the scarcity of the fuels in localised areas and their unavailability to weaker consumer groups will remain serious problems. With on-going trends towards strengthening market mechanisms and widening gaps in income distribution, an increasing number of traditional woodfuel consumers both in the domestic and small-scale industrial sector may become marginalized. Well-balanced and integrated forest policies can help to alleviate such problems.

Enhancement of supply in rural areas, where woodfuels is yet not a traded item, should be integrated into local farming and forestry management practices. Where woodfuel is mostly collected free of charge for subsistence, no prospect exists for its commercial production in short-run. In such a circumstance, local people's participation in sustainable production and utilization of woodfuel from locally available resources (mostly from existing natural forest and shrub/scrub and waste lands, and from existing depleted natural forest and shrub/scrub lands) should be encouraged. These resources possess the potential to supply additional woodfuel production if management systems which ensure protection from open cattle grazing and fires are introduced. Also, tree planting in community wastelands could contribute to the development of new supply sources, as village or community woodlots. Therefore continuation of the prevailing programme of social/community forestry, which primarily aims to promote participatory forestry development schemes, may be the most feasible low-cost strategy to meet the basic subsistence energy needs of the poor and small farming communities in rural areas.

CONCLUSIONS AND RECOMMENDATIONS

Conclusion

1. Wood energy is and will remain an important sub-

sector in all RWEDP member-countries. The consumption of wood and other biomass fuels will increase in the foreseeable future.

2. Non-forest land will continue to be the main source of woodfuels. Wood energy use is not and will not be a general or main cause of deforestation.
3. The prime area of concern is not the availability of woodfuels as such, but their distribution to people in need.
4. The weaker groups in society, particularly women and children, are the ones who suffer most from restricted access to woodfuel sources.
5. In Bangladesh and Pakistan, as well in Nepal to some extent, present national aggregated woodfuel consumption may exceed potential national supply. National woodfuel shortages may be aggravated by 2010.
6. In India, Sri Lanka, Thailand and Vietnam, aggregate national consumption in 1994 is not limited by aggregate potential supply, but this may be the case in 2010.
7. In most other RWEDP member-countries, residues from forests and crops represent and under-utilised potential to supplement woodfuel.
8. Localised woodfuel scarcities may occur in all countries.
9. The agricultural sector has a key role to play in supplementing woodfuels by enhancing woodfuel production on agricultural land.
10. The positive benefits of an integrated wood energy development strategy include; development of private, community woodlots in private and community owned lands which are currently not properly utilized; expansion of private, farm, and agro-forestry area; and support to conservation of soil, water and biodiversity.

11. For this integrated strategy to be successful, a number of issues need to be addressed which impinge upon the mandates of various sectors including the forestry sector.
12. In areas of woodfuel scarcity, other biomass fuels are likely to increase in importance as complementary sources of energy.
13. As a first approximation it can be stated that woodfuel use is carbon neutral, i.e. there is no net emission of carbon into the environment.
14. Thanks to woodfuel use in Asia, potential environmental costs amounting to at least 14 billion US $ in 1994, for recapturing CO_2 from the global environment were avoided. These will increase to 17.5 billion US$ in 2010.

Recommendations

1. The social, economic and environmental roles of woodfuels produced in both forest and non-forest areas should be recognised and woodfuels should be treated as an important sub-sector which needs to be developed.
2. Wood energy development should be integrated into rural energy supply strategies and pursued as a common task for all relevant sectors, e.g. agriculture, forestry, rural development, energy and industry sectors. Co-ordination among the sectors concerned should be strengthened.
3. Woodfuel should be seen as an important product in its own right rather than just as a by-product from agriculture land. Integrated woodfuel production on agriculture land should be promoted.
4. Current reforestation and afforestation efforts should be continued. Natural forest management with people's participation should get high priority in

areas where woodfuel is not (yet) a tradable commodity.

5. Prevailing rules and regulations which hamper wood energy development should be reviewed. These relate to land ownership and holding, tree tenure, tree planting and harvesting in private and community lands, transportation and trade of wood and related products produced by the private sector or local communities.
6. The selection of fast-growing tree species for wood energy crops, identification of appropriate provenance to match specific conditions, and improvement of the survival and growth rates of trees at degraded sites and waste lands, should be supported by further R&D.
7. Infrastructure should be developed further in areas where wood is already a traded item and where potential exists for supply enhancement to meet the existing and growing market demand.
8. The effective use of by-products and residues from wood industries, partly by converting them into modern wood energy, should be encouraged to reduce wood waste and supply additional fuels.
9. R&D for upgrading and combusting fuels from crop residues and other loose biomass should be promoted. Households as well as traditional industries should be encouraged to use them.
10. More key data on wood energy supply should be collected to support wood energy policies.
11. Wood energy databases should be established at regional, national and local levels. Private and public sector agencies related to wood energy development should be given access to information to support their activities.
12. Wood energy subjects should be integrated into the training curricula of relevant sectoral education and training programmes.

13. The priority within wood energy conservation programme should be the supply of convenient, healthy and attractive household stoves at affordable prices, so as to reach the maximum number of wood energy users.
14. The cost-effectiveness of wood energy development projects in Asia in terms of global CO_2 savings should be communicated to interested donor agencies.

ANNEXURE II

DR. ATMA RAM MEMORIAL LECTURE 1995

BIOMASS ENERGY: A MODERN AND ENVIRONMENTALLY FRIENDLY FUEL FOR THE FUTURE

Prof. David O. Hall & J.I. House

INTRODUCTION

Biomass is generally and wrongly regarded as a low status fuel, and rarely finds its way into the energy statistics when, in fact, it should be considered as a renewable equivalent to fossil fuels. It offers considerable flexibility of fuel supply due to the range and diversity of fuels which can be produced. Biomass can be burnt directly to produce electricity or heat, or it can be converted into solid, gaseous and liquid fuels using conversion technologies such as fermentation to produce alcohols, bacterial digestion to produce biogas, and gasification to produce a natural gas substitute. Industrial, agricultural and forest residues can be used as a biomass source, or energy crops such as trees and sugarcane can be grown specifically for conversion to energy.

Biomass is considered one of the key renewable energy resources of the future due to its large potential, economic viability and various social and environmental benefits. By 2050, biomass could provide 17% of the world's electricity and 38% of the world's direct fuel use (liquids, gas, heat) (Johansson *et al.*,1993). Presently, it is the fourth largest source of energy in the world, supplying about 13% of primary energy use (50 EJ/yr or 24 million barrels of oil a day, 1990 figures). Developing countries as a whole derive 38% of their energy from biomass, and in many it provides over 80% of total energy use in the form of traditional fuels, e.g. fuelwood, straw and dung. Since 90% of the world's population may reside in developing countries by 2050 biomass energy demand is likely to rise, even though per capita consumption may decline.

The traditional uses of biomass in developing countries—burning wood, agricultural residues and dung—are increasingly associated with the scarcity of hand-gathered wood, nutrient depletion and the problems of deforestation and land degradation. Such uses are often very inefficient (typically only 5-15% of energy is actually utilised) and can also give rise to harmful indoor air pollutants (Smith, 1993). The energy delivered is often less convenient to use compared to electricity, gas and kerosene, yet biomass is virtually "free" resource costing "only" labour to collect. Since poverty is likely to always exist there will be a continuing reliance on traditional biomass, so it is important to ensure that people have continued and sustainable access by not removing their land rights and other resources. Nonetheless, it is inevitable and desirable that the incomes of many will rise, and this will lead to a shift to more convenient forms of energy, such as gas, liquid and electricity, which can also be produced by modern biomass conversions.

If developing countries move towards using more coal and other fossil fuels, this will have serious effects on the global atmosphere. Similarly, charcoal is generally a very inefficient means of using biomass for households, increasing per capita wood use and contributing to deforestation. This is the real energy crisis of the future, how to encourage development while protecting the environment (Leach, 1988; Soussan, 1990), and how to reduce dependency on fossil fuels in industrialised countries without effecting standards of living. We believe that provision of modern, environmentally—

sustainable biomass energy system is part of the solution. Since biomass is the single most important energy resource in developing regions it is imperative that it be modernised in order to provide both the traditional energy needs and modern energy services (OTA,1992).

There is an enormous untapped biomass potential, particularly in improved utilisation of existing forest and other land resources, higher plant productivity and efficient conversion processes using advanced technologies. Therefore much more useful energy could be extracted from biomass than at present. Biomass can then form part of a matrix of fuel sources offering increased flexibility of fuel supply and energy security. This in turn could reduce dependency on oil imports which already crippled the economy of many countries. Bioenergy can be used at small and large scales in a decentralised manner bringing substantial benefits both to rural and urban areas.

Local (decentralised) and efficient production of modern biomass fuels will enable communities to remain self-sufficient in energy while promoting development. In particular this can bring substantial benefits to rural (and even urban) areas which don't usually have access to modern energy carriers. Growing biomass is a labour intensive actively and can create jobs in rural areas whilst providing energy to promote other rural industries, thereby helping to stem urban migration whilst providing convenient energy carriers. Job creation will also be of importance in industrialised countries, and growing biomass could provide an economically viable use for the agricultural land being taken out of production in "set-aside" schemes aimed at decreasing production of surplus agricultural commodities in Europe and North America.

A number of developed countries also use biomass quite substantially e.g. the USA derives 4% of its total energy from biomass (equivalent to 1.5 million barrels of oil per day and nearly as much as from nuclear power), Finland 18%, Sweden derives 14%, and Austria 13%. OECD Europe as a whole used 2 EJ of biomass energy in 1990, about 3.6% of total primary energy consumption, although this figure is probably an underestimate due to inadequate monitoring of rural and agroindustrial biomass energy use (ECN, 1994). Estimates of the potential of bioenergy in Europe in the next century range from 2 to 20 EJ/year (Johansson *et al.*, 1993; Hall, 1994).

The World Energy Council (WEC, 1993) broadly agrees with this calculated that 12% of 1990 energy use was from biomass; 930 million tonnes of oil equivalent (Mtoe) (10.5%,40 EJ) in the form of traditional uses and 121 Mtoe (1.4%, 5 EJ) from "modern biomass". They predict that under a current policy scenario, by 2020, modern uses will have doubled to 243 Mtoe (10 EJ), and traditional uses will have increased to 1323 Mtoe (56 EJ). Under an ecologically driven scenario, contributions could be 1060 Mtoe (45 EJ) from traditional uses and 561 (23 EJ) Mtoe from modern uses.

Biomass and the Greenhouse Effect

Plants, from minute algae through to large forests, absorb the sun's energy and carbon dioxide from the atmosphere which they store in their biomass. Burning and other conversions of plant biomass into a fuel source does not result in net carbon emissions since the biofuels will only release the amount of carbon they have absorbed during growth (providing production and harvesting is sustainable). If these biofuels are used instead of fossil fuels, carbon emissions from the displaced fossil fuels are avoided as well as other associated pollutants such as sulphur. Therefore, promoting biomass as a substitute for fossil fuels along with other renewable forms of energy, and energy efficiency could have a large impact on atmospheric CO_2 levels. Bioenergy can be sold and is competitive in many circumstances, therefore the net cost of offsetting CO_2 emissions could be near zero or even negative. Since industrialised countries have made a commitment to reduce CO_2, this could be a large incentive in the development of a modern biofuels industry. Biomass crops such as trees can also be planted on degraded, deforested and poor quality land helping to rehabilitate it and improve natural resource management (Hall *et al*, 1993; OTA, 1992; Bekkering, 1992).

2. BIOMASS SOURCES

Residues from forestry and agriculture are invaluable as an immediate and relatively cheap energy resource to provide the initial feedstock in the development of a bioenergy industry; this also offers an environmentally acceptable way of disposing of unwanted and often polluting wastes, but must encompass environmental sustainability. The ash from burners and effluent from digesters can be returned to the land as fertiliser. Table 2 shows that the energy content of potentially harvestable residues (the total residues produced that could potentially be collected) is about 93 EJ/yr worldwide. Assuming that only 25% of this is realistically recoverable, this could provide 7% of the world's energy use (Hall *et al*., 1993).

While residues will provide an important starting point for the bioenergy industry, the development large-scale energy production from biomass will probably rely in the future on specifically-grown energy crops such as sugarcane, rape seed, switch grass and trees (particularly short rotation forestry). Biomass productivities must be improved since they are generally low, being much less than 5 t (dry weight)/ha/yr for woody species without good management. It is now possible with good management, research, and planting of selected species and clones on appropriate soils to obtain 10 to 15 t/ha/yr in temperate areas and 15 to 25 t/ha/yr in tropical countries. Record yields of 40 t/ha/yr have been obtained with Eucalyptus in Brazil and Ethiopia. High yields are also feasible with herbaceous (non-woody) crops; for example, in Brazil, the average yield of sugarcane has risen from 47 to 65 t/ha (harvested weight) over the last 15 years while over 100 t/ha/yr are common in a number of areas such as Hawaii, South Africa and Queensland (Hall *et al.*, 1993).

Large-scale biomass crop production raises many questions about land availability and environmental impacts which must be addressed if bioenergy is to be acceptable to the public.

3. LAND AVAILABILITY

Land availability is perceived as a constraint to large scale production of biomass, however there are considerable areas potentially available. Large areas of surplus agricultural land in North America and Europe could become significant biomass producing areas. In the USA, farmers are paid not to farm about 10% of their land, and in the EC, 15% arable farmland is being "set-aside" (Howlett & Sargeant, 1993). Apart from over 30 Mha of cropland already set aside in the USA to reduce production or conserve land, another 43 Mha of croplands have high erosion rates and a further 43 Mha have "wetness" problems, which could be eased with a shift to various perennial energy crops. There are estimates that a further 60 Mha may be idled over the next 25 years (USDA, 1987).

In the European Union, at least 15-20 Mha of good agricultural land (an area of the size of England and Wales) is expected to be taken out of production by the year 2000/2010, and upto 50 Mha is predicted to be surplus to requirement in 12 EU countries by 2010 without even considering the potentially available and surplus land in E. Europe and the Former Soviet Union. If 20 Mha of land were used to plant trees, it would represent an annual sink of 120 MT C for the near future (assuming that over-dry biomass is about 50% carbon and the EC Biomass for Energy Programme productivity target of 12 t/ha is met). Alternatively, this area of land could provide 4.8 EJ/yr of biomass energy, displacing

120 MT of carbon emissions from coal, 96 MT from oil, or 67 MT of carbon emissions from natural gas (17% of total EC carbon emissions in 1991) (Hall & House, 1995a).

In tropical countries, there are large amounts of deforested and degraded lands that would benefit from the establishment of bioenergy plantations. For example, Trexler (1993) found that within 50 tropical countries, 67 Mha could be realistically converted to plantations over the next 60 years, more than 200 Mha could be regenerated and a further 63 Mha are available for agroforestry. Nilsson (1993) found that 265 Mha were available for reforestation and 84.5 Mha were available for agroforestry.

Assuming a productivity of 12 t (oven dry)/ha/yr, an energy content of oven-dry biomass of 20 GJ/t, and that recoverable residues (25% of potentially harvestable residues) are also used, this shows that 950 Mha would be needed to grow biomass energy crops to substitute for fossil fuels in industrialised areas, yet developing countries will require "only" 305 Mha. Therefore, on a global scale, there is enough land available to allow biomass to make a significant impact on atmospheric carbon levels and energy production. Developing countries could take economic advantage of their excess land by producing biomass fuels for export to industrialised regions, or by allowing the richer countries to pay for plantation establishment to offset their carbon emissions as is already heppening in trial schemes.

4. ENVIRONMENTAL ISSUES AND GUIDELINES

Environmental Sustainability

With the realization that biomass energy could become part of the modern energy economy on a large scale, there have been increasing concern as to the short and long term environmental effects of such a strategy. Fortunately, a number of environmental and biomass energy groups acknowledged some time ago that if biomass was to play an important role in future energy policy, then its production, conversion and use must be environmentally acceptable and also accepted by the public. These two factors are likely to be the most crucial constraints on future biomass developments and must be addressed in some detail in order for scenario builders to appreciate the opportunities and problems associated with biomass energy production. Several studies give for a more comprehensive coverage of environmental factors than what follows below (Beyea *et al.*, 1992; Evans, 1992; Pasztor & Kristoferson, 1990; Ranney, 1992; OTA, 1993; Shell/WWF, 1993; Gustafsson, 1994).

If plantations were to replace natural forests, there would be destructive environmental effects and negative effects on carbon inventories; this should be avoided. On the other hand, forestry projects established on degraded land or abandoned cropland can have numerous positive environmental effects. Reforestation/afforestation of degraded or agricultural land can improve soil structure, soil organic matter and nutrients, reduce run-off and increase soil water storage, increase local rainfall and modify local temperatures, increase biodiversity and preserve wildlife habitats, reduce pressure on natural forests, create windbreaks, and, of course, store carbon. With many energy crops, planting is far less frequent than for food crops: short rotation forestry crops are only replanted every 15 to 20 years; for perennial grasses with a three-year cutting cycle, replanting occurs perhaps once a decade. This means the land is disturbed far less frequently reducing erosion, nutrient loss and the need for herbicide application. Recent experience with the Conservation Reserve Program of the U.S. Department of Agriculture showed that the erosion rate declined 92% on 14 Mha of highly erodible U.S. cropland taken out of annual production and planted with perennial grasses and trees (Ranney, 1992). Furthermore, if the biomass is used to replace fossil fuel, it can offset many other environmental impacts such as acid rain from sulphur emissions.

There are wide concerns about the long-term biological health and stability of large plantations which rely on monocultures of selected clones, usually associated with high agrochemical inputs. If biomass energy crops were to replace monocultural food crops, or if they are established on degraded lands, in most cases the shift would be to a more biologically diverse landscape. Achieving sustainable biomass production while maintaining biological diversity may ultimately require a shift to polycultural strategies (e.g. mixed species in various planting configurations) which can be difficult to design and implement, but are capable of giving high yields. Polycultural establishment and management techniques therefore warrant high priority in research and development programmes for energy crops. Species selected for the plantations should be fast-growing and well-matched to the plantation site (e.g. for water requirements, seasonality, soil pH, salinity and nutrient balance, as well as susceptibility to herbivores, fire, disease, and pests). Consideration must also be given to the crop's potential impact on the surrounding habitat, with care taken to prevent aggressive exotic species from establishing themselves outside the plantation and displacing native flora. Exotic species are often more productive than indigenous species, but may be less well adapted to the local environment. Therefore, the use of clonal strategies could facilitate the incorporation of desirable characteristics while producing tree stands that are more likely to survive (Beyea *et al.*, 1992)

Care in species selection and good plantation design and management can be helpful in controlling pests and diseases, rendering the use of chemical pesticides unnecessary in all but special circumstances. Good plantation design includes: (i) areas set aside for native flora and fauna to harbour natural predators for pest control, and perhaps (ii) blocks characterized by different clones and/or species. If a pest attack breaks out in a block of one clone, a now common practice in well-managed plantations is to let the attack run its course and to let predators from the set-aside area help halt the pest outbreak. Some well managed plantations in Brazil now leave 20-30% of the area in a natural/undisturbed state. Establishing and maintaining natural reserves also enhances biodiversity, however, preserving biodiversity on a regional basis will require land-use planning in which natural forest patches are connected via a network of undisturbed corridors (riparian buffer zones, shelter belts, and hedgerows between fields), thus enabling species to migrate from the habitat to another (Hall *et al.*,1993).

Attention must be given to long-term soil fertility which also involves soil management. For short-rotation forestry, it will usually be desirable to leave leaves and twigs in the field, as nutrients tend to concentrate in these parts of the plant. Also, the mineral nutrients recovered as ash at the energy conversion facilities should be returned to the site. Species can be selected for efficiency of nutrient use; in addition, either selecting a nitrogen-fixing species or intercropping the primary crop with a N-fixing species can make the plantation self-sufficient in nitrogen. The promise of intercropping strategies is suggested by 10 year trials in Hawaii, where yields of 25 t/ha/y have been achieved without annual N fertilizer additions when Eucalyptus is interplanted with N-fixing Albizia trees (Wormald, 1992). Biodegradable agrochemicals should be used whenever possible, and their application should be carefully planned to match the needs of the plants. Furthermore, careful water management will reduce the risks of water pollution from run-off and make optimum use of rainfall in dry areas. Good fire management practices must also be included (Evans, 1992).

Environmental Guidelines

If biomass plantations are to play major roles in the global energy economy, strategies are needed for achieving and sustaining high yields over large areas and long periods. Despite the embryonic state of understanding of what is required, some general guidelines can be provided, drawing on the limited erxperience with plantations and much broader experiences with agriculture and forestry (Beyea *et al*., 1992)' The guidelines should address the main concerns of environmentalists and the public, while ensuring that the biomass growers are able to operate efficiently and profitably, and that social and other benefits are optimised. Such a balancing act needs careful consideration and planning if long term benefits are to accrue. Biodiversity and land use issues, technologies for production and conversion, and bioenergy development policies should all be included. The Shell/WWF (1993) "Guidelines" report sets out to establish "a general framework of principles within which more detailed guidelines may be developed and assessed". Most publications stress the need for ongoing, good quality monitoring and the practical feedback to bioenergy policy making. Environmental impacts of biomass production, residue removals and bioenergy use on soil, water and air quality and on habitats, cropping practices and greenhouse gas emissions all need to be fully understood if recommendations on best practice are to be made. The OTA (1993) background paper recommends that continuing R&D and environmental monitoring is necessary alongwith long term and large scale demonstration programmes.

COSTS

Most biomass energy technologies have not yet reached a stage where market forces alone can make the adoption of these technologies possible. One of the principal barriers to the commercialisation of all renewable energy technologies is that current energy markets mostly ignore the social and environmental costs and risks associated with conventional fuel use (Hohmeyer & Ottinger, 1994). Furthermore, conventional energy sources tend to receive large subsidies and support (Kosmo, 1987; Hubbard, 1991).

Therefore, renewable energy sources, which produce few or no external costs and have several positive external efects are systematically put at a disadvantage. Internalising external costs and benefits and re-allocating subsidies in a more equitable manner must become a priority for all renewables to be in a better ("level playing field") position to compete with fossil fuels. Some governments are trying to develop programmes to account for external costs such as taxes on emissions and incentives for cleaner fuels, but few schemes have yet been put into practice and most have met with strong opposition. In general, current emissions regulations and charges bear little relation to the real damage imposed as they tend to be political compromises.

Despite these distortions of the markets, many biomass energy schemes are already competitive (although regional and site-specific factors determining the cost of delivered biomass can alter pricing significantly). A recent study by the World Bank (Ahmed, 1994) examined biomass electricity and ethanol costs worldwide on a comparable basis from a total of 114 reports. In the case of electricity, it was concluded that "costs.. compare well with the costs of fossil-fired generation and even hydro generation in favourable situations; some are as low as 2-4 c/Wh... biomass gasification combined-cycle technologies show much promise for reducing costs for large-scale power generation...". Ethanol from biomass, with data mostly from the US and Brazil, show that the "decline in costs since the 1970s has been significant and is attributable to technology improvements and a shift toward cheaper crops". "The costs of ethanol were beginning to compare well with gasoline until the collapse of oil prices in the mid-1980s."

The Conference Proceedings (VTT, 1993; NREL, 1993; Hall *et al.*, 1994) and the reviews by Williams and Larson (Williams & Larson, 1993) are also good summaries of costs of biomass electricity and fuels and a comparison with other energy sources. Elliott (1993) concludes that the emerging technology of gasifier/turbines can give electricity at about 5 c/kWh and that the investment costs could be as low as $ 1300/kW after the 10th plant was installed. A recent report by the OECD (1994) has comprehensively synthesized and reanalysed data on biomass liquid fuels (ethanol, methanol, rape methyl ester) and electricity. The cost ratio of biofuels compared to gasoline or diesel varies from 3.4 to 7.8 (at EU feedstock prices, this ratio is lower with world feedstock prices), while the cost ratio for electricity from wood is 1.3 to 1.5 (6.4-9.5 c/kWh). The net energy ratio and CO_2 abatement potential are also more favourable for electricity from wood than for liquid biofuels.

The obvious question is whether these micro-economic comparisons justify shifting agricultural subsidies away from (surplus) food production toward biomass energy production or whether a more macroeconomic view incorporating environmental and social benefits is justified. A general conclusion of these economic analyses so far is that using woody biomass as an electricity feedstock results in the most favourable product costs. However, the use of starch and sugar crops may be a desirable short term use of biomass if the economic incentives are adequately structured.

6. EXAMPLES OF BIOENERGY USE

Solid Fuels Combustion

Biomass can be burnt to produce electricity in the same way as coal. In fact, it is becoming increasingly advocated to use coal and biomass fuels together in co-fired electricity generating systems. This is one way to reduce CO_2 emissions without the need to invest in new generating equipment. Sweden and Denmark incinerate 55 and 70% respectively of their municipal solid wastes, usually using the energy produced for district heating or combined heat and power (CHP) (Richards *et al.*, 1993). Normal electricity-generating power stations are only around 33% efficient because most of the heat produced is wasted. Combined heat and power (CHP) stations collect the waste heat and use it to heat the plant itself, or pump it to nearby industries and houses; this raises the efficiency to around 80%. Denmark has more than 50 straw-fired district heating plants that produce energy at a cost of $ 12/GJ, which is competitive with coal and oil-fired heating plants while providing an environmentally acceptable means of dealing with waste straw. They also have four CHP plants, with two under construction rated at 60 MW (Ravn-Jensen, 1992).

Liquid Biofuels

Bioethanol is becoming increasingly more prevalent in developing countries where raw materials are widely available and technologies are already demonstrated (Johansson *et al.*, 1993). Sugarcane is the world's largest source and the by-products and residues can be used as a convenient on-site electricity source, providing all the plant's requirements and excess to sell to the grid. Alcohols have favourable combustion characteristics, namely clean burning, high efficiency and high octane-rated performance (although a lower energy density) (Johansson *et al.*, 1993).

Brazil: In Brazil in 1989, over 12 billion litres of ethanol replaced about 200,000 barrels of imported oil a day. Almost five million automobiles now run on pure bioethanol and a further nine million run on a 20 to 22% blend of alcohol and gasoline (Goldemberg *et al.*, 1993) From 1976 to 1987, the total investment in the ProAlcool reached $ 6.97 billion and the total savings equivalent in imported gasoline was $ 12.48 billion (Lima, 1989). 700,000 direct jobs with perhaps 3 to 4 times this number of indirect jobs have been created (Goldemberg *et al.*, 1993). Economics could be further improved if sugarcane residues were to be fully utilised, they could provide for all on-site electricity needs, with excess being sold to the grid (Ogden *et al.*, 1990).

Europe: An EU Committee has set a target that upto 5% of the liquid fuel market could be supplied by biofuels by 2010 using an agricultural area of 7 Mha (European Communities Economic and Social Committee, 1992). Currently, commercial quantities of biodiesel are produced in Italy, France and Austria. One reason for the rising interest is the GATT set aside restrictions on the production of oilseeds and other crops, which does not apply to non-food uses such as the production of biofuels. The current cost per litre of biodiesel in the EC is US $ 0.39-0.46, which compares to US $ 0.23-0.31 per litre of mineral diesel. A draft directive has been proposed that would cut excise duty on biofuels to 10% of that charged on petrol and diesel. The EC Commissioner for taxes feels this is justified on environmental grounds and as a means of increasing energy security and protecting farmers' incomes (POST, 1993; Mauguin *et al.*, 1994).

Zimbabwe: Annual production of 40 M litres has been possible since 1983, except during the severe 1991/2 drought (Scurlock *et al.*, 1991). Production costs in 1988 were approximately $ 0.75/gallon (Steinglass, 1988) which indicates at least a break-even with gasoline imports. The ethanol plant at Triangle has operated successfully for almost a decade (Rosenchein, 1991), financed mainly by local capital, with home based technology required whenever possible rather than sophisticated equipment from abroad, thus allowing the use of low cost technology closely tailored to the industry's needs. All decisions concerning construction and running of the plant were made locally. The capital cost was only $ 6.4 million (1980 prices) —the lowest capital cost per litre for any ethanol plant in the world. Triangle offers an example of good use of relatively simple technology, local infrastructure and political commitment.

Biogas

Methane gas is produced naturally by the breakdown of organic matter. This process can be contained and controlled in biogas digesters, and the gas can be collected as an energy source. India has around 1 million household digesters, with a target of installing 12 million by 2,000. China has around 5 million, with a further 10,000 industrial digesters and over 800 biogas electric power stations. Denmark has ten centralised and ten single-farm biogas plants with an output of 0.5 PJ per annum (Woods & Hall, 1994).

Gasification

Gas can be produced by the thermochemical decomposition of biomass in an atmosphere that has a restricted supply of air. Millions of dollars have been spent on the R&D of thermochemical gasification of coal as a gas turbine feedstock, and much of this work is directly relevant to biomass integrated gasifier/gas turbines (BIG/GTs). Biomass is easier to gasify than coal because it is more reactive and has a very low sulphur content. It could produce electricity at a lower cost than large coal power plants with flue gas desulphurisation, nuclear power plants, and hydroelectric power plants. According to Williams and Larson (1993) efficiencies of 40% or more will be demonstrated in the mid 1990s, and by 2025 these could reach 57% using fuel-cell technologies being developed for coal. Gasifiers using wood and charcoal are again becoming commercially available, and research is being carried out on ways of gasifying other biomass fuels (such as residues) in some parts of the world (Foley & Barnard, 1993) The sugarcane industries that produce sugar and fuel ethanol are promising targets for near term-applications of BIG/GT technologies (Ogden *et al.*, 1990). Two gasification plants are coming on stream in Sweden and Finland, and a 30 MW_e plant is planned for Brazil. The EU have recently (October 1994) announced plans to construct three 8 to 10 MW_e plants using short rotation forestry as a feedstock.

India: Gasification has been the focus of attention in India because of its potential for large scale commercialisation (Paul & Mukunda, 1993). A detailed micro- and macro-analysis by Jain (1989) showed that the overall potential in terms of installed capacity could be as large as 10,000 to 20,000 MW by the year 2000, consisting of small scale decentralised installations for irrigation pumping and village electrification, as well as captive industrial power generation and grid-fed power from energy plantations. The ASTRA group in Bangalore have set up a wood gasifier in Hosahalli village (Ravindranath *et al.*, 1990). The wood gasifier provides gas to drive a 7 HP dual-fuel engine (80% biogas and 20% diesel) and 5 KVA 3-phase generator. An important aspect of this project is that the villagers are prepared to pay over twice as much for their electricity than from the grid (which is subsidised) because the supply is reliable and of a good quality, there are ancillary benefits (clean drinking water, flour mill etc.), and there has been an emergence of self reliance including the formation of a village management committee (Dutt & Ravindranath, 1993)

Other Examples of Bioenergy Use

Europe: Finland obtains 18% of its primary energy from biomass (5 MTOE in 1990). Over half of the large district heating plants use biofuels. Peat-fired combined heat and power plants are being introduced in some larger towns. One of the reasons for the success of the bioenergy industry is Government support. Currently, 10% of the total financing of energy research (US $ 2.5 million) goes to bioenergy. According to the Finnish Ministry of Trade and Industry, there are enough biomass resources available to allow for a doubling of their use for energy (VTT, 1992). The Swedish

Government has made a commitment to close all 12 nuclear power plants by 2010, that the four presently unexploited rivers will remain free of hydropower plants, and that atmospheric CO_2 will be stabilised at 1988 levels. Therefore Sweden now obtains 16% of its energy from biofuels, and even imports a small amount (NUTEK, 1992). Austria obtains about 13% of its primary energy from wood due to a political commitment including favourable legislation, capital grants, cheap finance and education. Farmers groups developing woodfuel-fired district heating schemes can receive grants covering up to 50% of the capital costs (Howes, 1992).

The UK only gets 1% of its primary energy from all renewables (a third of this is from biomass). The Government introduced the Non-Fossil Fuel Obligation (NFFO) to force the privatised electricity companies to purchase a certain quota of nuclear and renewably produced electricity, and use of biomass has increased by 20% over the last 5 years. Over 100 successful biomass schemes are now in operation, mainly using agro-industrial, household and sewage wastes, and growing energy coppice. In December 1994, the government awarded NFO (non-fossil fuel obligation) contracts for a further 627 MW_e, of which 444 MW_e will come from all types of biomass, including 19 MW_e incorporating gasification technology—the largest project was a jointly funded EU project which will use 2,600 ha of short rotation willow in an 8 MW_e plant. It is possible to provide 5% of the projected energy market in 2010 by establishing 5 million hectares of coppice, which is believed to be an achievable target (Seed, 1993). By 2010 there could be 5.5 Mha of "spare" agricultural land. According to the Energy Technology Support Unit (ETSU), every tonne of dry coppice could generate as much electricity as 650 kg of coal, and save the release of up to 500 kg of carbon. By 2005, energy crops could displace 18 MT of carbon emissions from coal, 10% of Britain's current annual output (Pearce, 1995) The UK is the World's third largest user of landfill gas with over 50 projects providing over 0.5 TWh/year out of an estimated potential of 5 TWh/year. The potential for using municipal solid wastes (MWS) is 15 TWh/yr (Richards *et al.*, 1993).

The USA: The USA currently obtains about 4% of its energy from biomass (Equivalent to 1.5 million barrels of oil a day, nearly as much as from nuclear power). The Electric Power Research Institute of the US (EPRI, 1993) feel that "Projections of emerging renewable energy markets in the first part of the 21st century consistently suggest that biomass energy systems will play a major role. . . The projections are based on the following facts: (1) vast availability of suitable land (tens of millions of acres) for growing energy crops, (2) environmental benefits (CO_2 neutral, low SO_2 and recyclable ash), and (3) favourable economics compared to coal-derived energy in the 1990s. Further, biomass plantations can help the economy in rural America."

Mauritius: Mauritius is dominated by the production of sugar, which represents around 88% of the cultivable area. Almost 60% of total energy requirements in Mauritius (excluding wood fuels primarily used for cooking) are met by bagasse-fired generation of power and steam in the sugarcane industry. This includes around 10% of Mauritius' electricity requirements. There is a potential for producing 10.2×10^6 GJ, using the by-products of the sugar industry and to a lesser extent solar energy. The Flacq United Estate Limited (FUEL), the largest sugar estate in Mauritius, installed a dual fuel, bagasse and coal furnace to produce electricity all-round the year. In 1989 the excess electricity exported to the grid reached 94×10^6 kWh (26×10^6 from bagasse) representing about 16% of the country's total requirements (Baguant, 1990). The government has set up a Bagasse Energy Development Programme (BEDP) which aims to increase cane residue-derived electricity production from the present level of 70 GWh to about 120 GWh. This will exploit about 56% of the total potential from bagasse, but due to increased electricity demand, bagasse is only expected to provide about 9% of total electricity production by the year 2000 (GEF, 1992).

Northeast Brazil - Potential: Over 90% of all electricity produced in Brazil, and virtually all that produced in the Northeast, is hydroelectric. However, the hydroelectric potential will inevitably be exhausted at a reasonable cost and new sources must be found. 50 Mha of land sub-optimal for agriculture could produce 12.6 EJ/yr (total energy use in the Northeast is about 1.1 EJ). The present biomass energy production potential from the area of sugarcane planted in 1989 is estimated to be 174 PJ/yr. If tops and leaves (barbojo) are also used, the bioenergy available could be increased by up to 75%. The energy potentially available from other agricultural residues in the Northeast is estimated to be about 145 PJ/yr. Carpentieri *et al.* (1992) have compared scenarios for obtaining energy from biomass or hydro power. For the biomass scenario average unit investment costs would be 25% lower, the total required capital investment would be half as much, average electricity production costs will be lower, and marginal production costs will be substantially lower. The Brazilian Government is promising to introduce a new policy across Brazil that will mandate the state-controlled electricity utilities to enter into long-term contracts to buy co-generated power. This will encourage further growth of biomass electricity production from residues (Carpentieri *et al.*, 1992). The development and financing of a 30 MW gasifier plant is progressing rapidly and is due for construction by 1994. India's primary energy consumption in 1992 was in the range of 356 to 425 Mtoe (14.9 to 17.8 EJ) depending on whether a low or mean estimates is taken for biomass use in the domestic sector. Of this, biomass supplies 36 to 46%. Total primary energy use per capita is in the range 17.7 to 21.1 GJ/yr which is very low compared to the OECD countries (200 GJ/capita/

yr). As population increases and levels of energy consumption per capita improve, energy use is increasing at a rapid rate. Between 1981 and 1991 the use of commercial energy rose by 4.5%, and the use of traditional biomass energy use in rural areas is also increasing (Ravindranath & Hall, 1995). The 75% of the population that resides in rural areas have largely not benefited from modern commercial energy or energy planning which usually focuses on increasing the total use of energy (mainly in the form of fossil fuels) per capita rather than delivering the energy services required by a housewife, farmer, transporter or an entrepreneur to promote development.

In their recent book, Ravindranath & Hall (1995) have analyzed biomass energy use in India and the potential for modern biomass to provide energy for rural development and agroindustries. This study is an attempt to focus on the large bioenergy potential which exists in any country but at present is very poorly utilised in supplying energy services in parallel with conventional energy systems. The study examined the possibility of providing quality modern fuels to fulfil the needs in any typical village situation such as gaseous and liquid fuels for cooking in all homes; electricity for lighting in all homes; community services in villages (pumping drinking water, street lighting and flour milling); and electricity for all shaft-power requiring activities (pumping irrigation water and agro-processing).

The technologies for decentralised electricity generation from biomass and biogas for cooking are available for different scales of operations, though significant scope exists for improvements in efficiency of conversion, use of multiple feedstock, and user friendliness. However, liquid fuel technology systems (using woody feedstocks) at the scales feasible for India, and high efficiency megawatt-scale producer gas systems are not yet available. Research and development efforts are continuing and efficient and viable bioenergy technologies will soon become available at competitive prices. Meanwhile it is recommended to establish rural bioenergy centres where all the village energy requirements can be considered in an integrated way so that bioenergy options can be matched to need. Bioenergy programmes must become an integral part of the overall rural development programme through genuine empowerment and participation of local communities.

The analysis showed that the energy requirements of all services necessary to promote development and quality of life in rural areas could be met using modern bioenergy options. Additionally this would make a significant contribution to alleviating the stress on forests and the national electricity grid, thus paving the way for sustainable development and reduced dependence on oil imports. The basic premise of the analysis was to provide quality fuels to meet the energy services required to promote the quality of life of the dominant rural population. This could be achieved through technologies that are sustainable and which promote local control, self-reliance and equity. The new bioenergy technologies satisfy these criteria and further provide other tangible benefits such as local employment, land reclamation, and reduction in emissions of greenhouse gases. The study also shows the unique opportunity that bioenergy provides for local, national and global benefits without any of the serious conflicts which often characterise conventional energy options. Land availability is not a constraint even in a densely populated country like India.

Land and Sustainable Biomass Supply

The analysis identified a need for dedicated energy plantations in order to generate about 122 MT of woody biomass annually for decentralised electricity production. Since the existing forests are already depleted and there is need for increasing food production to meet the needs of a rising population, the question arises as to where this biomass will be grown.

Land is not a constraint to meet all the biomass needs of a growing population: food, fodder, timber, and fuel (for bioenergy). The area under cropland has been stable since 1970 in India, even though the population has been growing at 2% annually. Crop productivity and cropping intensity are low, giving rise to a large potential to double or even triple food production without any increase in area under cropping. Large areas of farm and communal lands are degraded and not suitable for profitable cropping. Bioenergy could provide the necessary incentives to reclaim currently degraded lands, which have little possibility of being rehabilitated via conventional afforestation or fodder development programmes. Even with a low energy conversion efficiency of 18%, only 15 to 30 Mha (at productivities of 8 to 4 t/ha/yr respectively) are adequate for meeting biomass energy needs for 98% of rural settlements. A productivity of 8t/ha/yr or more is feasible and sustainable with appropriate seed quality, species mix, silvicultural practices, and soil and water conservation measures. The area of land required would decrease with potential increases in efficiency of energy conservation as well as plantation productivity, although increases in productivity must be sustainable, and guidelines may be appropriate.

Energy plantations with appropriate species mixes and density could also provide other products such as grass for livestock, leaves for biogas, and other valuable tree products (fruits, seeds, etc.). The inclusion of high value tree crops yielding fruit, flower and industrial feedstocks could increase the profitability of bioenergy programmes. Residues and dung are also widely available in rural villages and are a free resource thus can be easily converted into energy. Thus bioenergy programmes could become part of a rural development package involving reclamation of degraded watersheds, local employment generation, promotion of self reliance, and improving the quality of life especially for women and the rural poor.

It is interesting to examine India's deforestation record (Ravindranath & Hall, 1994). The recent FAO assessment of deforestation revealed an acceleration of tropical deforestation between 1981 and 1990, yet India has succeeded in significantly halting deforestation since 1980, successfully stabilising its forest area (and in turn the standing stock of carbon) and conserving its forest resources. This has occurred despite population growth and the large dependence of rural communities on forest resources. Conservationist traditions of many local communities, local and national pressure groups, progressive legislation (such as social forestry) are largely responsible for the decline in deforestation. Following on from this, a further argument in the case of bioenergy is that with the provision of efficient biogas and liquid fuels, the fuelwood requirement for cooking would become redundant, and this biomass could instead be used as a feedstock for modern conversion. Currently 181 MT of woody biomass is used annually in the rural domestic sector for cooking.

Large Bioenergy Systems

Megawatt-scale biomass electricity generation and liquid fuel systems could be of significant economic benefit to farmers who own degraded lands which could be dedicated to biomass production. They could also lead to reduced oil imports and replacement of coal in power generation. Since not all existing degraded lands will be required for village-scale small, decentralised schemes, surplus land may be available. Farmers could set-up megawatt-scale electricity generation units and supply feedstock in a similar way to sugar cooperatives, where growers in a given area set-up a sugar mill to process their cane. A potential of 16,000 to 32,000 MW of installed capacity is feasible by dedicating only 5% of India's geographic area. For a 1 MW system, biomass feedstock would come from a radius of only 8 km, at a productivity of 8t/ha/yr, using only 5% of the geographic area for energy plantations, and assuming only 18% conversion efficiency.

The motivation for the farmers are profits and employment from unused land. Sugar mills also have a large potential for cogeneration of bioenergy from residues (bagasse) which has been recognised by the government to be at least 3,500 MW. This option needs to be perused on a priority basis since bagasse is already available in a centralised location as a by-product which is used inefficiently.

Institutional and Financial Policy Options

Even though the potential for bioenergy exists there are financial and institutional barriers, though none are insurmountable given public acceptability and political commitment. Technical potential of bioenergy plus the significant environmental and social benefits are unlikely to lead to any large-scale shifts to bioenergy. Even provision of financial subsidies will not promote the bioenergy options in the absence of a whole range of policy measures. For example, no amount of subsidy in investment cost for a megawatt-sized electricity plant will help if there are restrictions on the sale of electricity to buyers at economic prices, or if it has to face state-owned utilities providing free or highly subsidised electricity. In their book, Ravindranath and Hall discuss in detail the barriers and the problems in large-scale commercialisation or dissemination of bioenergy, the roles of different actors involved, and they suggest a set of guidelines. However, they accept that one must be realistic—energy policies are more often based on short-term political considerations rather than rational financial or environmental criteria.

In developing countries, energy procurement, production, distribution and pricing is largely controlled by state utilities, and high subsidies on electricity and petroleum fuels are prevalent. There is a growing realisation that price subsidies to infrastructure almost always benefit the non-poor disproportionately (World Bank, 1994). A bioenergy entrepreneur, farmer, or village community will face many obstacles in the production and transportation of biomass. The removal of existing restrictions on energy productions and sale, the empowerment of rural communities (to which the government is committed on paper), along with promotion of private initiatives and appropriate financial incentives may act as a catalyst for decentralised bioenergy production.

Bioenergy as an energy source is also certainly not without its problems, like all other energy sources. Undoubtedly the energy plans of any country will consist of a diversity of energy sources and technologies including renewable as well as non-renewable sources and, consequently, bioenergy will only be a part of the overall supply mix.

A Bioresource Strategy for India

Indian bioresources were recently analyzed at a Biomass Users Network (BUN) conference in Bangalore (Reddy, 1994) which found that despite an abundance of policies, there was an almost complete lack of energy goals and strategies. The conclusions were that the goals for bioresources must be compatible with, and contribute to, the goal of the Indian socio-economic system which is sustainable development, i.e. an equitable, empowering and environmentally sound process of growth. The bioresources sub-system should, therefore, be an instrument of sustainable development, contributing to the satisfaction of basic needs, starting from those of the neediest. It should strengthen self-reliance at the local, state and national levels. It should also be in harmony with the environment to ensure long-term sustainability of the development process. And, needless to add, it has to be economically efficient so that it results in economic growth. In order to advance sustainable development, a necessary condition is an increase and eventually maximisation of the sustainable utilisation of bioresources.

The analysis highlighted the importance of ensuring sustainability in the production and utilisation of bioresources; increasing, and eventually maximising, the role of bioresources; protecting and improving the resource base for bioresource production and utilisation; and optimising the mix of bioresources outputs—food, fuel, fodder, fibre, fertiliser, forest products (including medicine) and financially rewarding (cash) crops—all of which are important. In order for a successful bioresource strategy to be implemented, the following factors were found to be important: emphasising employment and income generation; making bioresource production and utilisation economically viable; ensuring people's participation and control with special emphasis on women and the poor; emphasising decentralised settlement-based development of bioresources; coordinating and restructuring and/or reorienting existing institutions dealing with bioresources; designing new institutions to deal with bioresources; formulating and implementing conducive policies; arranging financing mechanisms; emphasising research and development; and creating and strengthening appropriate technical capacity.

8. CARBON MITIGATION POTENTIAL

Biomass as a Carbon Store or Energy Source?

There are two main ways in which biomass can reduce atmospheric CO_2 levels. The first is by simply growing biomass to absorb CO_2 emissions from the atmosphere and act as a carbon store; the second is to use this biomass as an energy source substituting for fossil fuels and offsetting emissions from this source. There are many factors to consider, and ultimately the best method will depend on local circumstances.

It is sometimes argues that to grow enough trees as an energy source to displace a given amount of fossil fuels requires more land than to grow enough trees to absorb and store (sequester) the carbon emissions from the same fossil fuels. However, carbon sequestration by trees usually ceases at forest maturity whereas periodic harvesting of biomass for energy allows land to be used (and therefore fossil fuels to be replaced) indefinitely (Hall *et al*., 1990; Hall *et al*.,1992)

In terms of economics, while the production of bioenergy is more costly than simply growing trees as standing stock, there are revenues from the sale of the energy produced which can offset the costs of plantation establishment. Since biomass derived electricity and liquid fuels can be produced competitively in certain circumstances, the net cost of offsetting CO_2 emissions by substitution could be near zero or negative (Hall *et al*., 1990). In addition to this economic advantage, there are also the numerous environmental and social benefits associated with bioenergy provision that are mentioned elsewhere in this paper.

Marland & Marland (1992) compared the two strategies of carbon storage or fossil fuel substitution by modelling carbon flows. They found that, depending on the assumptions made, growing trees for carbon storage may be more appropriate on low productivity land (or indeed where the biomass cannot be practically harvested). However, they conclude that "where high productivity can be expected, the most effective strategy is to manage the forest for a harvestable crop and to use the harvest for maximum efficiency either for long-lived products or to substitute for fossil fuels." Using biomass to substitute for fossil fuels is likely to be a more advantageous and appropriate method for reducing atmospheric CO_2 level using available land, as long as environmental sustainability is ensured (Hall *et al*.,1993). Nevertheless, a combination of carbon storage and fossil fuel substitution may be realistic in circumstances where harvesting the biomass at frequent intervals is difficult or financially unrewarding.

9. FUTURE BIOMASS OPTIONS: PHOTOBIOLOGY

Future biological methods for atmospheric CO_2 reduction include laboratory based research on using microalgae and using photobiochemistry to fix CO_2. Microalgae can be grown efficiently in photobioreactors and tanks or ponds, to remove CO_2 directly from power station flue gases (Hall & Rao, 1990; Nishikawa *et al.*, 1992). The biomass produced could provide a source of energy, chemicals and food, while wastes can be converted or recycled into useful byproducts (e.g. fertilizer). These photobiological systems have a considerably higher photosynthetic efficiency than conventional biomass systems, and do not require high quality land or water so would not compete with agriculture and forestry. A Japanese team have already managed to grow microalgae at a photosynthetic efficiency of 4.1% (Nishikawa *et al.*, 1992). Assuming, for example, it would be possible to achieve a 5% conversion efficiency, with average solar energy in the Mediterranean region of about 200 W/m^2 or 6.3 GJ/m^2/yr, 18 Mha of land in this area would theoretically be required to fulfil all western Europe's energy requirements; this is only 11.6% if all the land in France, Greece, Italy, Portugal & Spain (Hall *et al.*, 1992).

Researchers in the UK (Coghlan, 1993) have used a "Biocoil" (TM) reactor (transparent coiled tube) for growing green algae (Chlorella) in a nutrient medium bubbled with a CO_2 stream. The reactor is illuminated artificially and/or using natural sunlight, and algae are drawn off, dried and milled, and fed into a duel-fuel engine in which 95% of the fuel is algae; the waste CO_2 is recycled back to the "Biocoil". This prototype system produces 25kW of power, and researchers estimate the costs of electricity from this process would be around 3 pence/kWh. Preliminary work by Watanabe & Hall (1995), growing the cyanobacterium Spirulina in a similar "Biocoil" reactor with 4% CO_2 in a gas stream, has shown that about 4% efficiencies (conversion of photosynthetically active radiation, 400-700 nm, to dry weight of cells) are feasible. Work on optimising the energy efficiency of the overall system continues.

Other uses of free-living and immobilised microalgae include the production of hydrogen in photobioreactors which can be used as a fuel source (Markov *et al.*, 1995). Although these technologies are still experimental, there is an unquestionable potential in the long term and industrial companies as well as academic researchers are already working on such approaches.

10. CONCLUSIONS

Bioenergy production and use must be modernised both to provide an environmentally sound fuel source that can provide for the rising demand for commercial energies, and as a means of reducing atmospheric carbon dioxide levels. Developing countries have been particularly instrumental in pushing modernisation of biomass for energy as they are so heavily dependent on it, and likely to remain so. Both developing and industrialised countries are now realising they can benefit environmentally and economically. Ancillary benefits of biomass production include restoration of degraded lands, energy security, foreign exchange savings, job creation and provision of electricity and other forms of energy to rural areas, thereby helping to promote development (Woods & Hall, 1994).

Of the major alternatives to reduce atmospheric CO_2 levels, and to solve future energy problems, biomass options are not perfect, but they are among the most promising and environmentally acceptable (Rubin *et al.*, 1992; Alpert *et al.*, 1992). However, they are only part of an overall strategy that should include energy efficiency, other alternative fuels, and emission control technologies.

There are large areas of degraded land available for revegetation in the topics as well as set-aside land in industrial countries. However, if large-scale programmes are to be successful there must be policy reforms such as to encourage environmental sustainability (including the establishment of environmental guidelines), improved infrastructure and planning, ongoing research and long term monitoring and a large financial commitment made at an international level. In particular, the efficiency of biomass production and use must be improved.

The largest problem faced by bioenergy is probably that of market penetration as almost all other problems can be avoided with careful planning and management, and local commitment. A balanced system of taxes and subsidies needs to be put in place to allow biomass to compete fairly in the energy market. Subsidies to fossil fuels should be reduced or those to renewable fuels increased so they are at a similar level, and the environmental and social costs ("externalities") of fossil fuels should be included in their pricing. Many biomass energy projects are already proving to be economically viable despite the currently low price of oil and subsidies to many energy forms. It is crucial that the perceptions of bioenergy by the public, industry and governments be improved. In this way, the various constraints may be overcome and entrepreneurial opportunities created. In particular, success will require that there must be local involvement, multiple benefits, sustainability and flexibility.

INDEX